21世纪高等学校规划教材 | 计算机科学与技术

Java Web开发基础
——从Servlet到JSP（第2版）

王斐 祝开艳 主编

清华大学出版社
北京

内容简介

本书用精简的篇幅讲解了Java Web开发需要的基础知识。从Servlet到JSP，再到设计模式的应用，带领读者由浅入深地掌握Java Web开发的基本技巧。最后采用迭代法实现了一个架构合理的MIS系统，并使用DAO设计模式和数据库连接池技术，让读者在迭代的过程中感受分层架构的优点。

本书适合作为计算机科学与技术专业、软件工程专业及相关专业的本科教材，也适合对Java Web开发感兴趣的读者自学使用。

图书在版编目(CIP)数据

Java Web开发基础：从Servlet到JSP/王斐，祝开艳主编. —2版. —北京：清华大学出版社，2019（2024.7重印）
（21世纪高等学校规划教材·计算机科学与技术）
ISBN 978-7-302-53000-8

Ⅰ. ①J… Ⅱ. ①王… ②祝… Ⅲ. ①JAVA语言—程序设计—高等学校—教材 Ⅳ. ①TP312.8

中国版本图书馆CIP数据核字(2019)第093924号

责任编辑：刘向威
封面设计：傅瑞学
责任校对：胡伟民
责任印制：杨 艳

出版发行：清华大学出版社
网　址：https://www.tup.com.cn，https://www.wqxuetang.com
地　址：北京清华大学学研大厦A座　**邮　编**：100084
社 总 机：010-83470000　**邮　购**：010-62786544
投稿与读者服务：010-62776969，c-service@tup.tsinghua.edu.cn
质量反馈：010-62772015，zhiliang@tup.tsinghua.edu.cn
课件下载：https://www.tup.com.cn，010-83470236
印 装 者：三河市人民印务有限公司
经　销：全国新华书店
开　本：185mm×260mm　**印　张**：17.75　**字　数**：447千字
版　次：2014年10月第1版　2019年8月第2版　**印　次**：2024年7月第6次印刷
印　数：8001～9500
定　价：49.00元

产品编号：081898-01

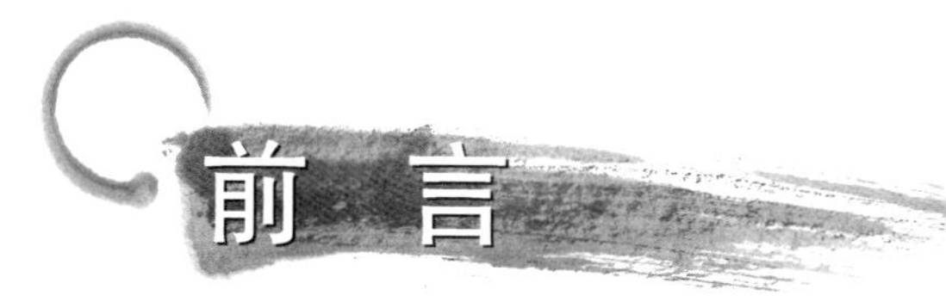

前言

本书是《Java Web 开发基础——从 Servlet 到 JSP》(ISBN:9787302373223)的第 2 版,继续保留原教材侧重实践教学、兼顾理论基础的特点。调整了部分章节的内容,进一步压缩了纯理论内容,只保留必要的部分。采用思维导图的形式对每章的内容进行小结,引导读者建立起清晰的 Web 应用程序开发脉络。

移动互联网方兴未艾,但传统的 Web 应用程序也没有就此消亡。各种 Web 开发框架层出不穷。有道是"万事开头难",在学习 Web 应用程序开发的过程中,很多人都有由入门到放弃的经历。原因就在于 Web 应用程序涉及的知识是综合性的,涵盖了多方面的知识。

本书按照技术发展的脉络,从介绍 Servlet 出发,到 JSP,再到设计模式,为读者将来学习各种 Web 开发框架打好基础。本书不求完备,在有些知识点上甚至是一带而过。书中所涉及的知识点如果是开发中必须解决的,那就做详细介绍;如果是暂时用不到或者多数情况下用不到的,就略讲或者完全不讲,留待感兴趣的同学自行钻研。本书内容组织上是一个逐步提高的体系,通过不断对代码进行改进,慢慢引入设计模式来改善代码质量。

本书在内容组织上,分成了五部分,如图 0-1 所示:

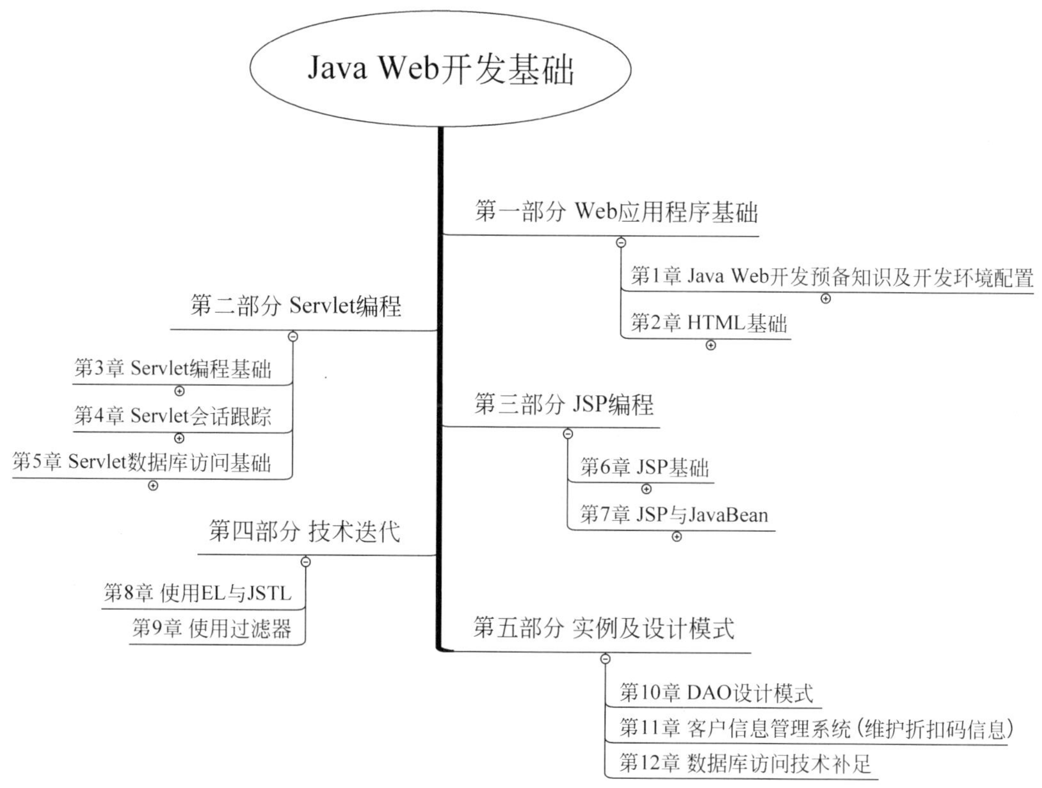

图 0-1 本书内容结构图

第一部分为 Web 应用程序基础，包括第 1 章、第 2 章。

第二部分为 Servlet 编程，包括第 3 章、第 4 章、第 5 章。

第三部分为 JSP 编程，包括第 6 章、第 7 章。

第四部分为技术迭代，包括第 8 章、第 9 章。

第五部分为实例及设计模式，包括第 10 章、第 11 章、第 12 章。

第 1～7 章由大连海洋大学信息工程学院祝开艳编写；第 8～12 章由大连交通大学软件学院王斐编写。

本书的主要特色是：

(1) 以实用为主，贴近开发。从开发中来，到教学中去(先讲企业开发的基础知识，逐步引入企业级开发常用的设计模式，案例设计更适合教学)。

(2) 不是语法书，不求大而全，但求小而精(从 Java Web 开发技术的演进出发，以 Servlet 始，到 Servlet、JSP、Java Bean 三者的综合应用终，与大多数的教材不同的是，对于已经逐步弱化的 JSP 部分，并没有如语法书般展开，而是有所取舍，重点讲述了开发中避无可避的问题)。

(3) 以问题为导向，在解决问题中推进(每个知识点都是围绕着问题的解决展开，每解决一个问题，就在知识深度上更进一步)。

(4) 每章结束时都以思维导图的形式做了总结(本书为了排版的需要，采用了逻辑图的结构)。

本书更适合高校开设 JSP 类课程时使用，也可以作为 Java EE 框架课程的先修课程。适合学时比较少，学校机房配置条件比较有限，在 Java 教学时未引入集成 IDE 的学校使用。它同时也适合自学的读者，因为本书选择的 IDE 基本上避免了自学时可能遇到的开发环境的困扰。

需要注意的是：

(1) 如果你想找一本语法书，随手查查，事无巨细，那么这本书不适合你。

(2) 如果你是一个开发高手，想要提高，那么这本书不适合你。

(3) 如果你是一个新人，想要自学 Java Web 开发，那么这本书很适合你。

(4) 如果你对 Java 学得不好，担心做不了 Java Web 开发，那么这本书很适合你。

(5) 如果你觉得自己英语不好，面对着英文的编程界面无比恐惧，那么这本书很适合你。

(6) 如果你想用很短的时间学会编写一个 Java Web 站点，并希望将来学习各种神奇的框架技术，那么这本书很适合你。

最后，不能免俗的是，作者水平有限，如果书中有错漏、疏忽的地方，还请大家批评指正。

编　者

2019 年 2 月

目　录

第1章 Java Web开发预备知识及开发环境配置

学习目标：

通过本章的学习，你应该：

- 理解 Web 应用程序、网站、网页的概念
- 理解静态网站与动态网站的概念
- 理解 Web 前端与 Web 后端的概念
- 理解开发 Web 应用程序需要解决的 3 个基本问题
- 理解 C/S 架构与 B/S 架构
- 搭建开发运行环境
- 熟悉 NetBeans IDE 的界面
- 掌握 NetBeans IDE 的基本用法

1.1 Web 应用程序基本概念

1. Web 应用程序

狭义地讲，Web 应用程序指的就是各种各样的网站(website)，网站由一系列的网页(webpage)构成。网站有静态网站和动态网站之分。

2. 静态网站与动态网站

静态网站：所有的网页，相同的站点，相同的时间，不同的人来访问，呈现出的内容都一样。这很像看电视，只要是相同的频道，相同的时间，不同的人看到的电视节目都一样。

动态网站：相同的站点，相同的时间，不同人访问，呈现出的内容是可以不一样的。例如各种需要登录才能访问的网站，相同的站点，相同的时间，不同的人登录之后看到的内容是不同的。

3. Web 前端与 Web 后端

在开发网站时，随着开发规模越来越大，开发人员之间也有了一定的分工。可以大致分成前端工程师、后端工程师、全栈工程师。其中全栈工程师覆盖了前端和后端。前端工程师、后端工程师、全栈工程师所侧重的技术栈各不相同。

Web 前端指的是客户看得见的部分，负责更好地和客户交互，做信息的呈现与提交。

从事前端工作的工作人员，一般被称为网页设计师(web designer)。技术栈主要涉及JS、CSS、DIV、HTML、JQuery等。

Web后端指的是服务器端的工作，负责具体的业务逻辑，一般会涉及数据库的操作，典型的操作是对数据的增、删、改、查。从事后端工作的工作人员，一般被称为网页开发人员(web developer)。从事后端开发的Java程序员，目前技术栈主要涉及Java、SQL、Spring MVC、MyBatis等。

总而言之：前端负责"貌美如花"，后端负责"增删改查"。

4. 开发Web应用程序需要解决的3个基本问题

开发Web应用程序，可以采用的技术方案很多，使用的开发语言也各不相同。但是无论采用哪种方案，都要解决3个基本问题：

(1) 如何将信息呈现给用户以及如何引导用户提交信息。

(2) 在服务器端如何获取及处理用户提交的信息。

(3) 在服务器端如何与数据库交互。

5. 桌面应用程序、网络应用程序、Web应用程序

在没有计算机网络之前，应用程序主要运行于单机环境，突出的特征就是软件需要先安装再运行，这就是桌面应用程序。在单机运行的桌面应用程序中，用户提交信息和处理业务逻辑都在本地。

有了计算机网络之后，就衍生出了网络应用程序。早期的E-mail、FTP应用均属此列，突出的特点是通常会有一个客户端、一个服务器。用户信息的处理一部分在客户端，一部分在服务器。这种架构被称为C/S(Client/Server)架构。除了E-mail、FTP等应用，很多的管理信息系统也采用了C/S架构，这类网络应用程序一般把数据存储在服务器上，这样数据可以供客户端共享。

C/S架构的软件有一些缺点是无法克服的，所以在C/S架构的软件基础上又产生了B/S(Browser/Server)架构的Web应用程序。

备注：也有人把Web应用程序称为网络应用程序。

6. C/S架构与B/S架构

1) C/S架构

C/S中文译为"客户机/服务器"，C/S架构如图1.1所示。

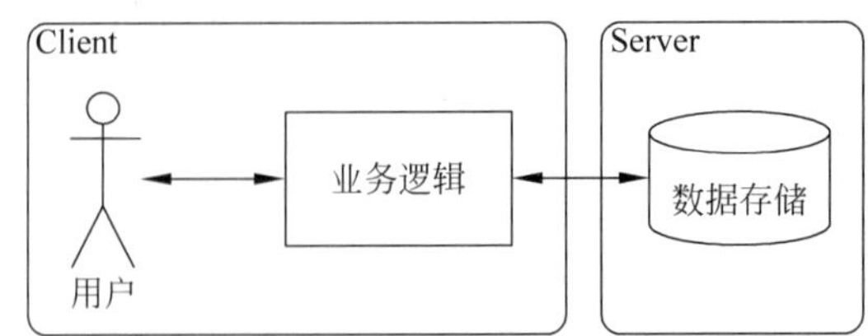

图1.1 C/S架构

在 C/S 架构中，因为客户的业务逻辑集中在客户端，因此它还有一个形象的名字叫作"胖客户端"，而这里的服务器严格地讲应该被称为数据库服务器，因为除了提供基础的数据库服务，服务器没有其他的功能。

2）B/S 架构

B/S 中文译为"浏览器/服务器"，B/S 架构如图 1.2 所示。

在 B/S 架构中，因为业务逻辑和数据操作都集中在服务器端，所以它也有一个形象的名字叫作"瘦客户端"，这里的服务器既包含应用服务器，也包含数据库服务器。

想要进一步了解 C/S 与 B/S 架构知识，请看本章末扩展阅读 1。

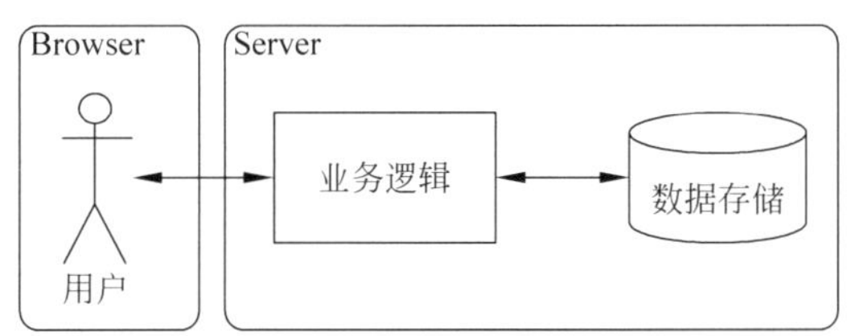

图 1.2　B/S 架构

1.2　Java Web 开发环境的搭建及相关介绍

本书采用 NetBeans IDE 作为开发环境。

Java Web 开发涉及的软件如下：

第一是要支持 Java 语言，需要下载一个 JDK。

第二是编写、编译管理 Java 代码及其他代码的集成开发环境，这里选择 NetBeans IDE。

第三是 Web 容器/服务器，本书选择 GlassFish Server/Tomcat（两个服务器的安装文件都集成在 NetBeans IDE 安装包中，安装过程中可以自由选择），因为后续例子与 GlassFish Server 有关联，采用本服务器会避免很多不必要的麻烦，所以强烈建议安装此服务器。

第四是数据库服务器，本书选择 Java DB（如果选择 GlassFish Server 作为服务器，那么 Java DB 默认已经同时安装好了）。

总结：需要下载一个 JDK 安装文件、一个 NetBeans IDE 安装文件。

1. 下载、安装 JDK

在搜索引擎中搜索 JDK，在搜索结果中选择官方网站，或者直接访问网址 https://www.oracle.com/technetwork/java/javase/downloads/index.html 均可到达 JDK 的下载页面。在该页面上单击左侧导航中的 Java SE 链接，然后选择 Downloads 标签页里的图标，如图 1.3 所示。

在图 1.3 页面滚动鼠标滚轮，选择 Java SE 8u181 版本 JDK（本书选择 Java SE 8u181，以便于与 NetBeans IDE 版本匹配）。单击矩形框处 DOWNLOAD 链接，如图 1.4 所示。

图 1.3　下载 JDK 示意图 1

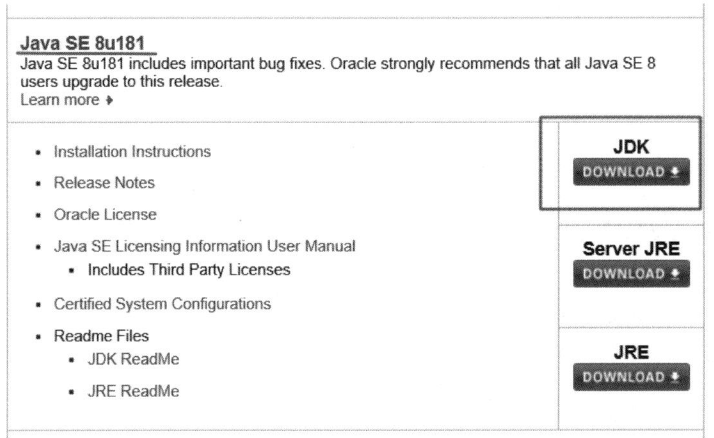

图 1.4　下载 JDK 示意图 2

在出现的页面上选择 Accept License Agreement，再单击自己需要的版本即可顺利下载 JDK（本书示例选择的是 Windows x64 版本，如图 1.5 所示）。

JDK 的安装比较简单，双击安装包，在安装向导的引导下，连续确认就可以了，所以不再截图示意，建议初学者选择默认安装路径。

2. 下载、安装 NetBeans IDE

在图 1.5 中，右侧选项卡中有直接打开 NetBeans IDE 下载页面的超级链接，或访问网址 https://netbeans.org/downloads/index.html，均可打开 NetBeans IDE 的下载页面。在该页面选择 Java EE 版本（矩形框部分）即可下载，如图 1.6 所示。

在 NetBeans IDE 的安装过程中，有几个地方是可以选择安装的。

双击安装包，首先系统会要求选择应用服务器，如图 1.7 所示。这个安装包里面默认包括了两个服务器，一个是 GlassFish Server，另一个是 Apache Tomcat。其实对于本书内容两个服务器任选其一即可，本书示例采用 GlassFish Server，所以选择安装 GlassFish Server。

单击“下一步”按钮，出现许可证协议界面，选择“我接受许可证协议中的条款”复选框，单击“下一步”按钮，如图 1.8 所示。

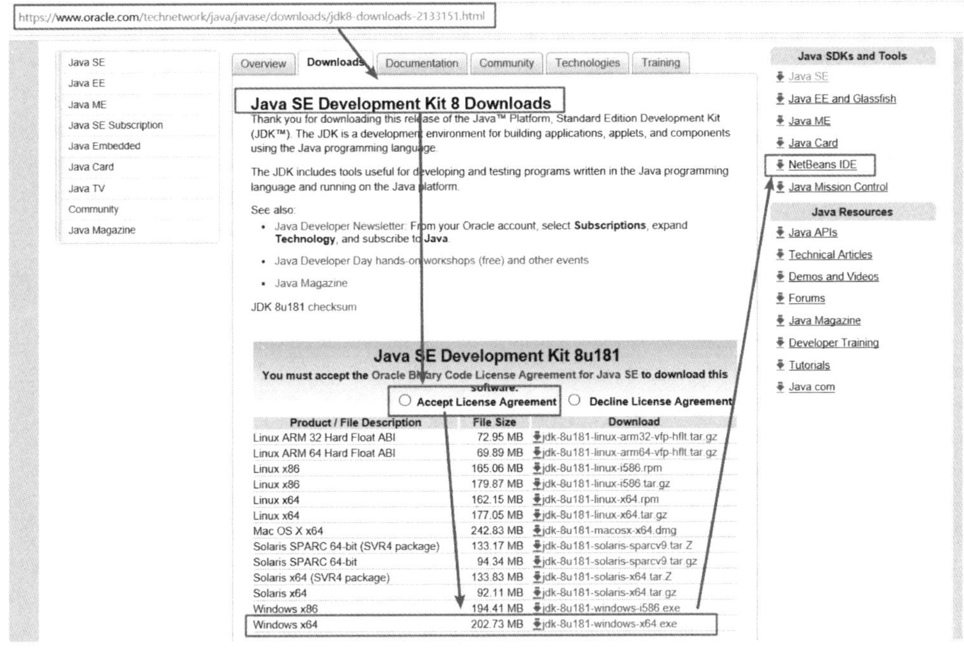

图 1.5　下载 JDK 示意图 3

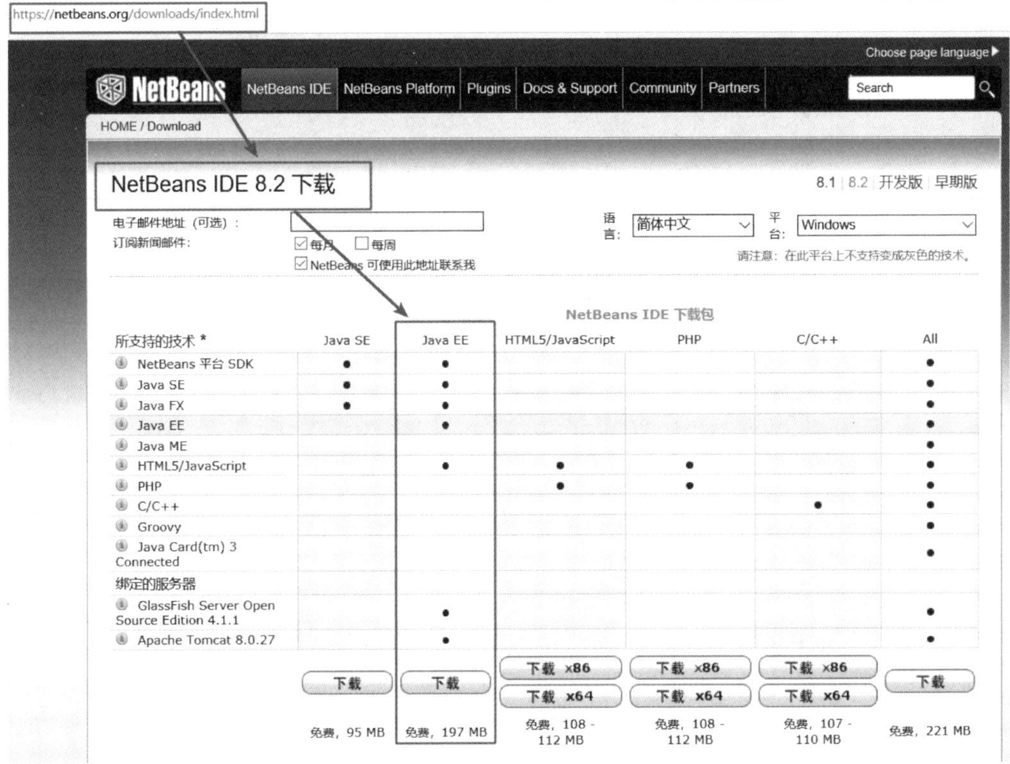

图 1.6　NetBeans IDE 下载示意图

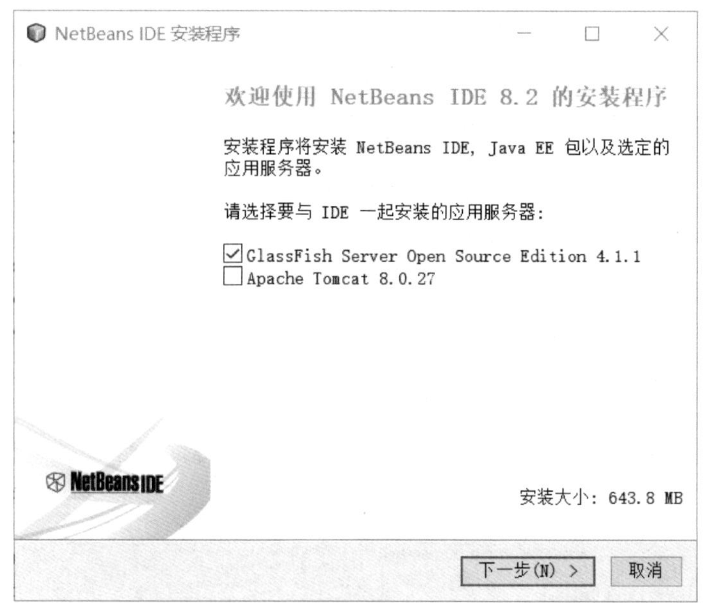

图 1.7　安装 NetBeans IDE 并选择安装 GlassFish Server

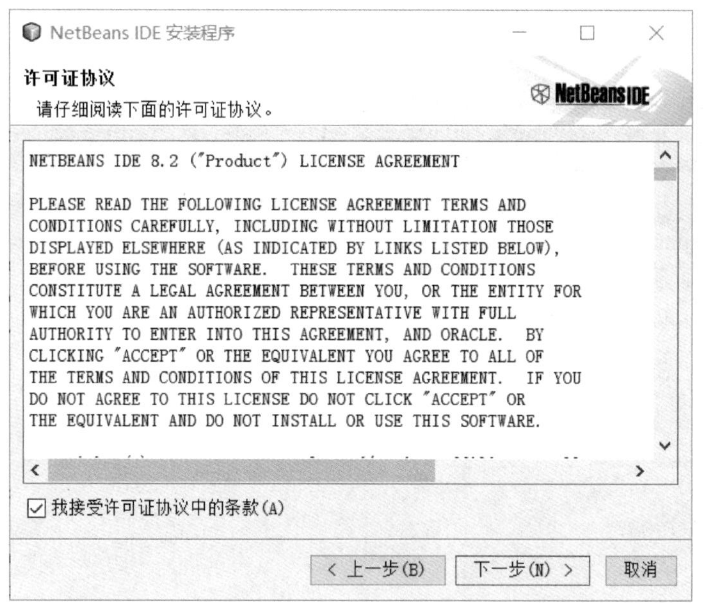

图 1.8　许可证协议选择

选择 NetBeans IDE 安装文件夹和选择用于 NetBeans IDE 的 JDK。

建议选择默认值,单击"下一步"按钮,如图 1.9 所示。

根据提示选择 GlassFish 安装路径及服务器使用的 JDK,单击"下一步"按钮,如图 1.10 所示。

根据提示启动安装操作,单击"安装"按钮,如图 1.11 所示。

安装完成后,单击"完成"按钮,安装过程结束,如图 1.12 所示。至此,本书中需要用到

图 1.9 选择 NetBeans IDE 安装文件夹及对应的 JDK

NetBeans IDE 安装程序
GlassFish 4.1.1 安装
选择安装文件夹。
NetBeans IDE
将 GlassFish 安装到(I):
C:\Program Files\glassfish-4.1.1
浏览(R)...
用于 GlassFish 应用服务器的 JDK(tm)(J):
C:\Program Files\Java\jdk1.8.0_181
浏览(O)...
< 上一步(B)
下一步(N) >
取消

图 1.10 选择 GlassFish 安装文件夹及使用的 JDK

的开发环境搭建完毕。

3. 使用 NetBeans IDE 创建 Java 应用程序

下面通过一个简单的示例演示如何创建 Java 控制台应用程序。

目标：学习使用 NetBeans IDE 创建 Java 控制台应用程序。

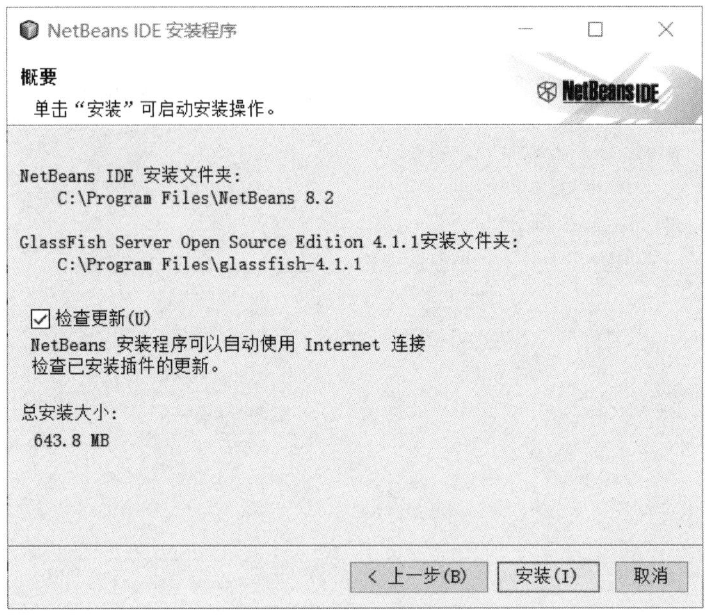

图 1.11　启动安装操作

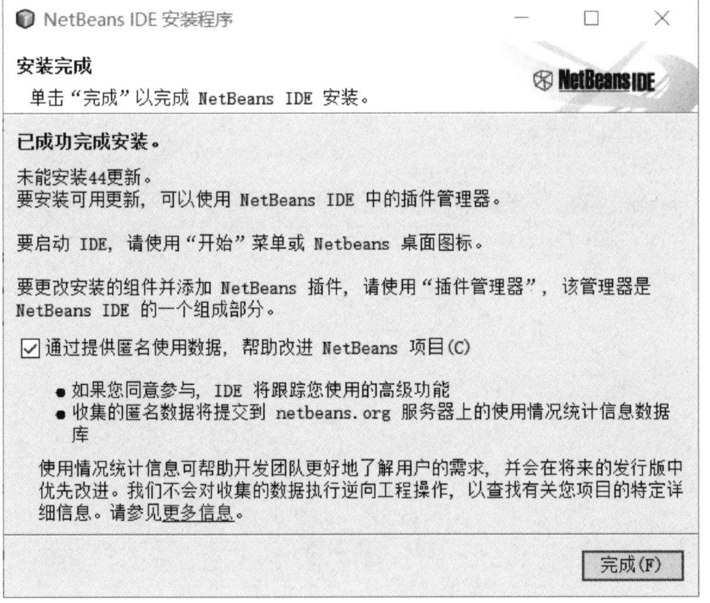

图 1.12　安装完成

工程名：eg0101。

运行 NetBeans IDE,启动界面如图 1.13 所示。

在菜单栏中选择“文件”→“新建项目”命令,在“新建项目”对话框的“类别”中选择 Java 选项,在右边的“项目”中选择“Java 应用程序”,如图 1.14 所示,然后单击“下一步”按钮。

设置项目名称为 eg0101,设置项目位置为 C:\NetBeansProjects,暂时不创建主类。单

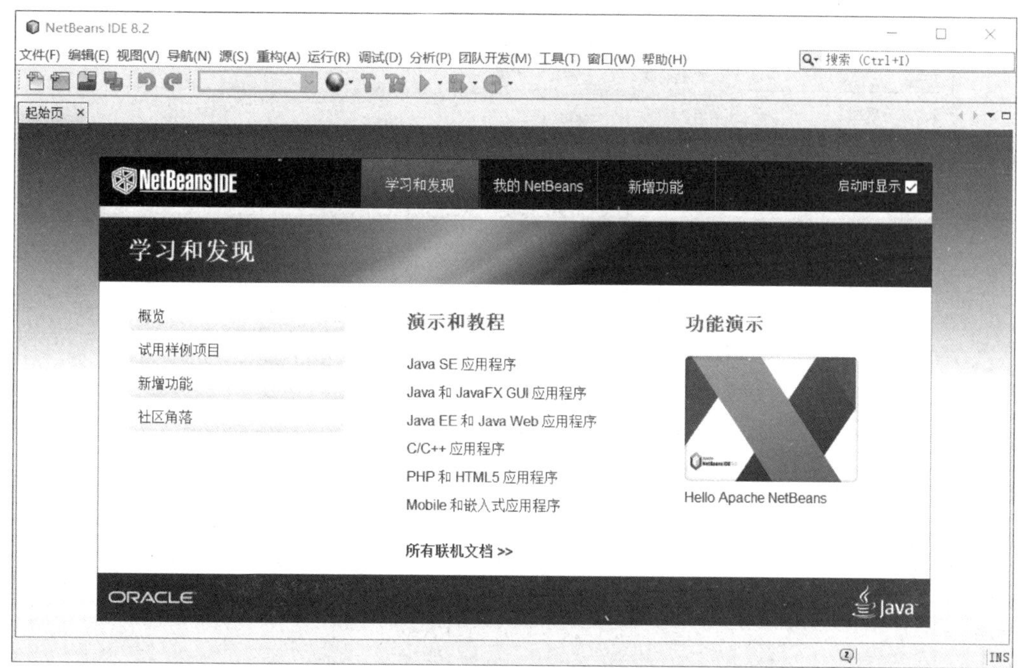

图 1.13　NetBeans IDE 启动界面

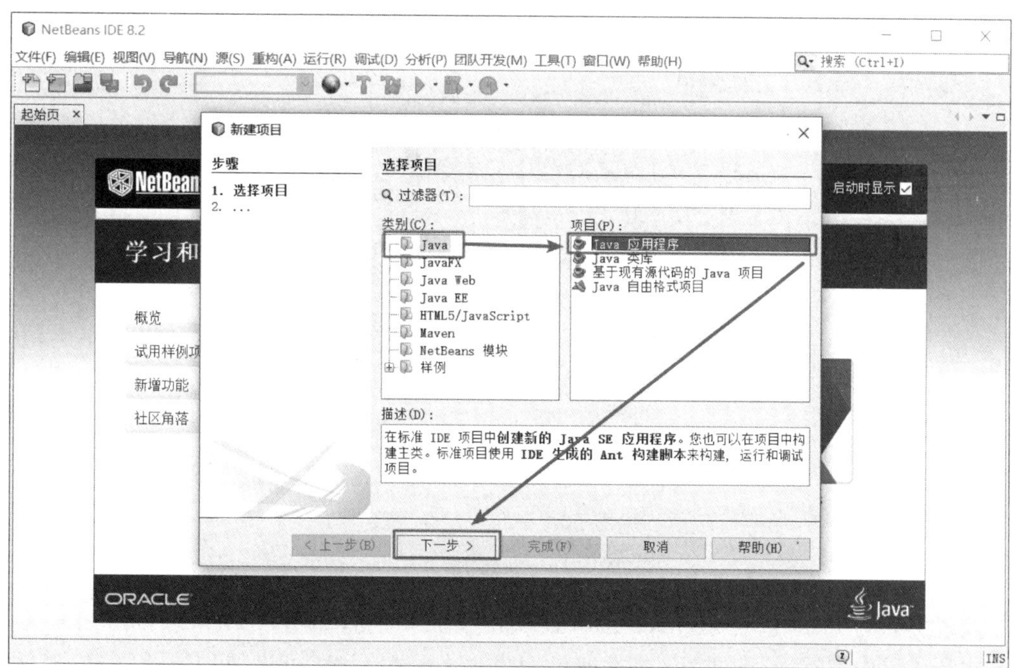

图 1.14　使用 NetBeans IDE 创建 Java 应用程序

击“完成”按钮，如图 1.15 所示。

在新建项目 eg0101 中，右击“默认包”选项，选择“新建”→“Java 类”命令，如图 1.16 所示。

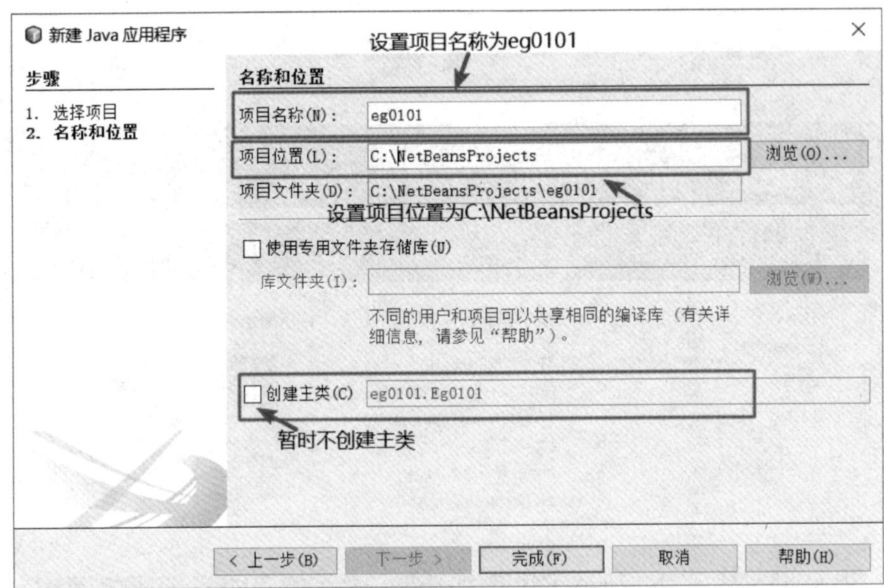

图 1.15 设置项目参数

图 1.16 新建 Java 类

设置类名为 Run，包名为 cn.edu.djtu，单击"完成"按钮，如图 1.17 所示。

在类代码中编写 main 方法，该类即可变成主类。编写时，可以使用快捷方式，先输入 psvm，然后按 Tab 键，可直接生成 main 方法，如图 1.18 所示。

在 main 方法中，编写控制台输出语句。编写时，可以使用快捷方式，先输入 sout，然后按 Tab 键，可直接生成输出语句，如图 1.19 所示。

选中 Run.java 类，如图 1.20 中 1 所示，或者在代码空白处，如图 1.20 中 2 所示，右击，

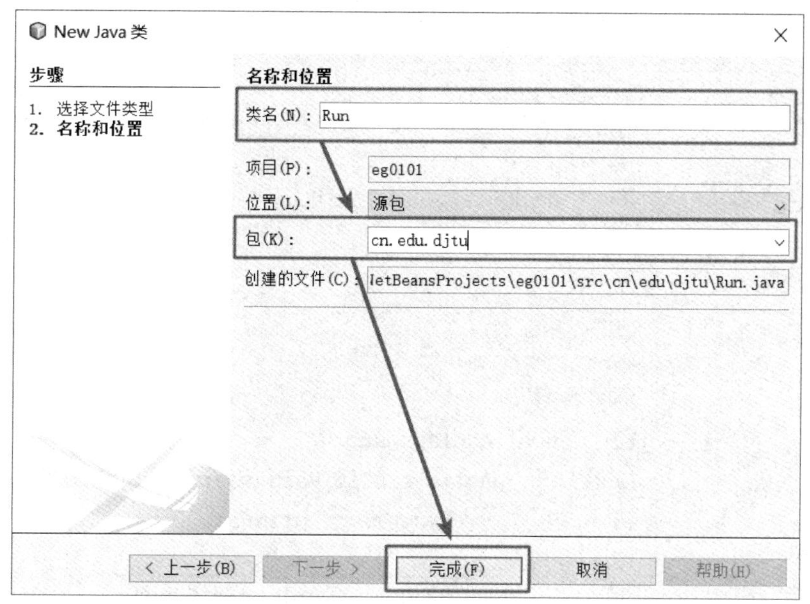

图 1.17 设置类参数

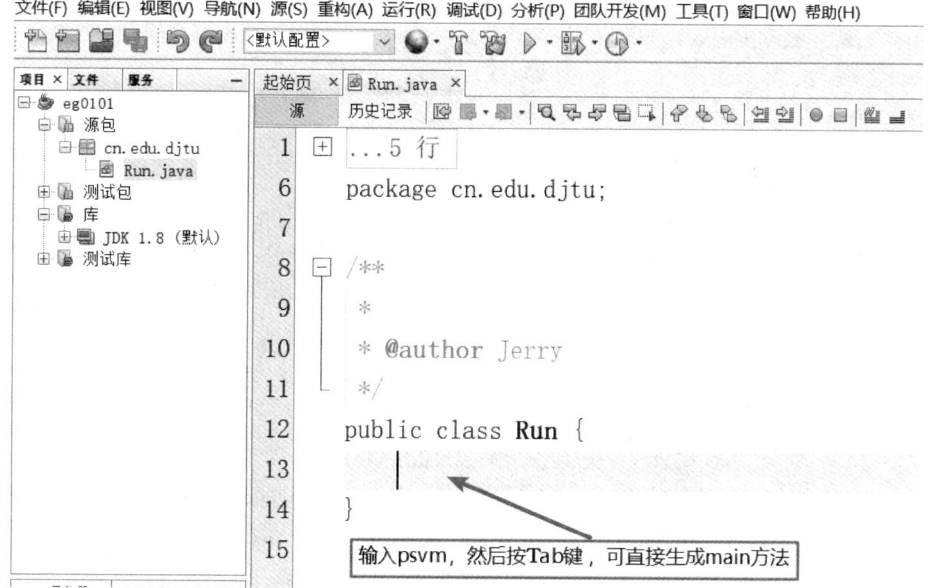

图 1.18 编写 main 方法

然后选择"运行文件"命令即可运行该主类，或者单击"运行"按钮，如图 1.20 中 3 所示，也可运行该类。输出结果将显示在代码下方的输出窗口处，如图 1.20 中 4 所示。

源代码默认字号较小，可以在菜单栏中选择"工具"→"选项"命令，在弹出的选项对话框中选择"字体和颜色"选项，设置成自己喜欢的样式，如图 1.21 所示。

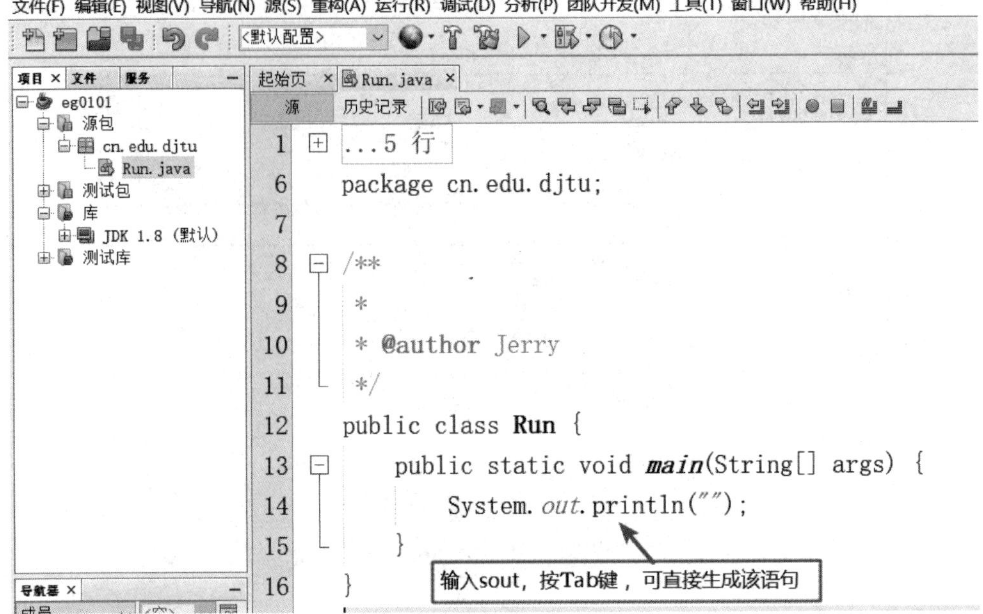

图 1.19 编写控制台输出语句

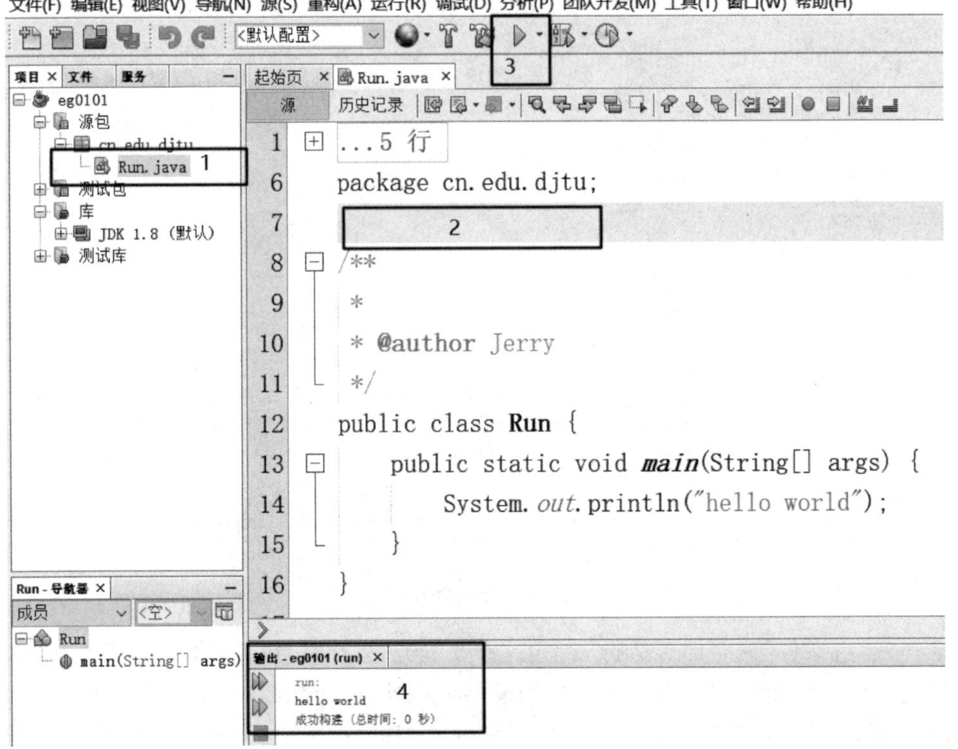

图 1.20 运行主类

图 1.21　设置字体和颜色

1.3　本章回顾

本章的主要内容分成两部分，一部分是 Java Web 开发的预备知识，另一部分是开发环境的配置。

预备知识主要涉及的基本概念如下：

(1) Web 应用程序。

(2) 静态网站与动态网站。

(3) Web 前端与 Web 后端。

(4) 开发 Web 应用程序需要解决的 3 个基本问题。

(5) 桌面应用程序、网络应用程序和 Web 应用程序。

(6) C/S 架构与 B/S 架构。

开发环境主要包括 JDK、NetBeans IDE 和 Glass Fish Server。

本章内容结构图如图 1.22 所示。

扩展阅读 1：

1) C/S 架构的优点与缺点

优点：

(1) 它能充分发挥客户端计算机的处理能力，客户端将用户请求处理后再提交给服务器，同时将服务器提供的数据处理后再以某种方式显示给客户，所以一个突出的优点就是客

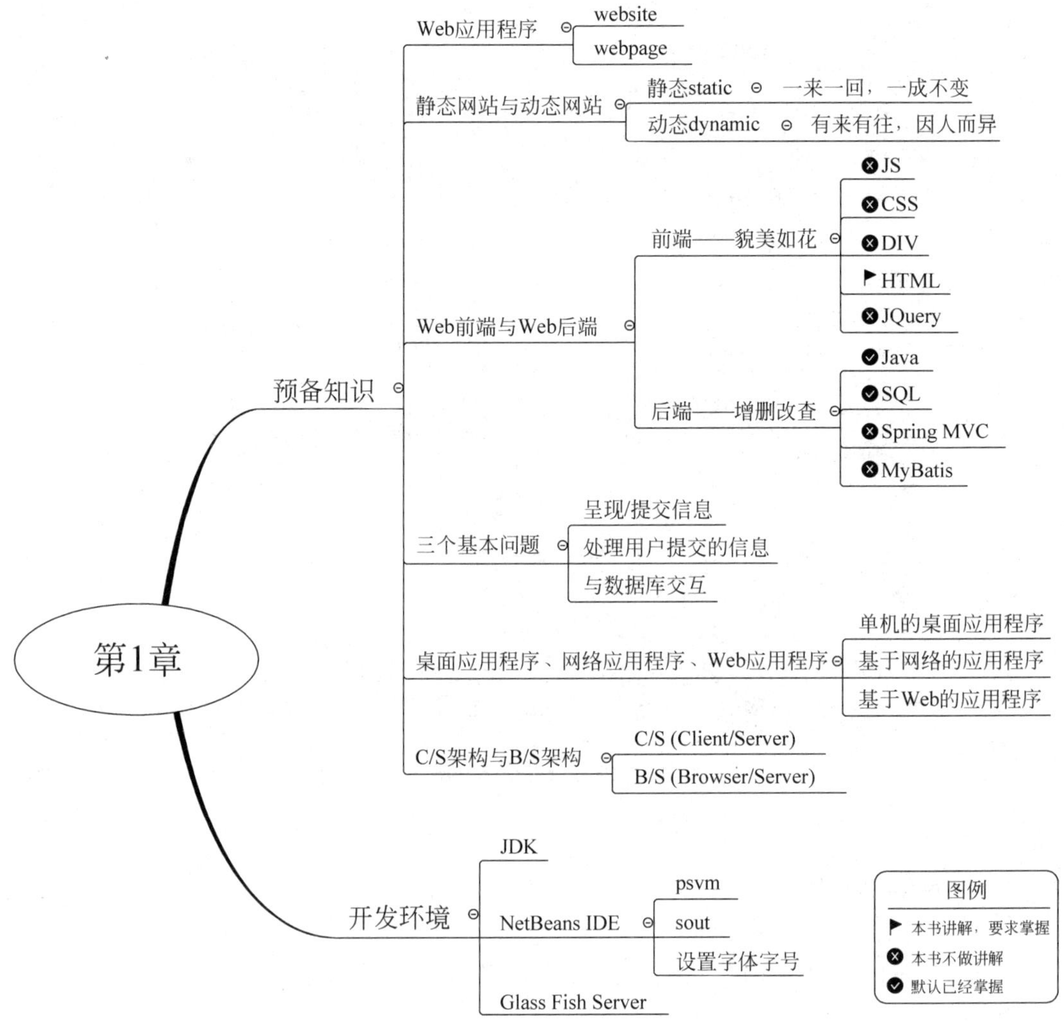

图 1.22 第 1 章内容结构图

户端响应速度快。

(2) 由于程序员在开发单机应用程序的过程中积累了大量可借鉴的经验，加之有高效的开发工具支持，所以开发效率很高，因此 C/S 架构在网络应用程序开发早期大行其道。即使在 B/S 架构非常流行的今天，C/S 架构依然有其用武之地，例如腾讯公司的 QQ、暴雪公司的魔兽世界等。

缺点：

(1) 客户端需要安装专用的客户端软件，在 C/S 架构下，全部业务逻辑的处理都集中在客户端，一旦应用的需求发生变化即业务逻辑发生变化，则客户端和服务器端的应用程序都需要进行修改。服务器端应用程序修改基本上是可控的，一是服务器数量少，二是主要涉及数据存储，改动量较小。但是，新的客户端应用程序必须重新分发给所有的用户端，这一工作量比较大，在以前网络传输速率不太理想的情况下，这一点的成本比较高。尽管现在网络传输速率已经有所改善，通过网络可以进行客户端程序的升级，但是从用户体验上看，并不建议采取强制用户升级的手段，而是将选择权交给用户，这就面临着客户端可能存在不一致的状况。

(2) 客户端应用程序往往是基于某一操作系统编写的，对客户端的操作系统一般也会有限制，针对 Windows 操作系统开发的客户端，通常并不适用于 Linux 或 UNIX，即使同是 Windows 系列的操作系统，在 Windows XP 下能运行的，也未必一定能在运行 Windows 7 操作系统下。

此外，随着互联网的飞速发展，移动办公和分布式办公越来越普及，通常需要系统具有扩展性，尽量少受客户端的限制，尤其是在应对客户输入较少、业务逻辑相对简单的场合。正是因为以上原因，在 C/S 架构的基础上又产生了 B/S 架构。

2) B/S 架构的优点与缺点

优点：

B/S 架构克服了 C/S 架构的一部分缺点，首先消除了客户端的差异，只要有浏览器就可以，无须另外安装客户端。一旦应用的需求发生变化即业务逻辑发生变化，只须更新服务器端的应用程序即可确保所有客户使用相同的新版本。其次也不受操作系统的限制，因为目前浏览器已经成为各个操作系统的标准配置，任何一个操作系统下都有多个浏览器产品可供选择。

缺点：

B/S 架构的缺点主要在于增加了服务器端的负担，因为业务逻辑集中在服务器端，服务器需要对每一个用户的每一次请求做出响应，同时又要对数据库进行维护，工作量较大。

扩展阅读 2：

J2EE 与 JavaEE

Sun 公司在 1998 年发表 JDK1.2 版本的时候，使用了新名称 Java 2 Platform，即"Java 2 平台"，修改后的 JDK 称为 Java 2 Platform Software Developing Kit，即 J2SDK，并分为标准版(Standard Edition，J2SE)、企业版(Enterprise Edition，J2EE)、微型版(Micro Edition，J2ME)。J2EE 便由此诞生。J2EE 是 Sun 公司为企业级应用推出的标准平台。

随着 Java 技术的发展，J2EE 平台得到了迅速的发展，成为 Java 语言中最活跃的体系之一。现如今，J2EE 不仅是指一种标准平台，它更多地表达着一种软件架构和设计思想。

2005 年 6 月，Java One 大会召开，Sun 公司公开 Java SE 6 版本。此时，Java 的各种版本已经更名并取消其中的数字"2"：J2EE 更名为 Java EE，J2SE 更名为 Java SE，J2ME 更名为 Java ME。Java 平台共分为 3 个主要版本 Java EE、Java SE 和 Java ME。

Java EE 是由一系列技术标准所组成的平台，在本书的课程体系中，主要用到 JDBC、JSP、JSTL 和 Servlet 这 4 种技术。JDBC 是指 Java 数据库联接(Java Database Connectivity)，JSP 是指 Java 服务器页面(Java Server Pages)，JSTL 是指 Java 服务器页面标准标签库(Java Server Pages Standard Tag Library)，Servlet 是指 Java 小服务程序(Java Servlet AP)。

1.4 课后习题

1. 什么是 Web 应用程序？
2. 什么是静态网站？什么是动态网站？试比较两者的区别。

3. 什么是 Web 前端？什么是 Web 后端？
4. 开发 Web 应用程序需要解决的 3 个基本问题是什么？
5. 什么是桌面应用程序？什么是网络应用程序？什么是 Web 应用程序？
6. 什么是 C/S 架构？什么是 B/S 架构？
7. 分别下载并安装合适版本的 JDK 和 NetBeans IDE。
8. 如何快捷地输入 System.out.println("")；？

HTML基础

学习目标：

通过本章的学习，你应该：

- 理解 HTML 在 Java Web 开发技术体系中的位置
- 掌握 HTML 的语法格式
- 掌握常用的 HTML 标签
- 掌握在 NetBeans 中编写 HTML 页面的技能

第 1 章提到开发 Web 应用程序，需要解决 3 个基本问题：

(1) 如何将信息呈现给用户以及如何引导用户提交信息。

(2) 在服务器端如何获取及处理用户提交的信息。

(3) 在服务器端如何与数据库交互。

本章探讨如何解决第一个问题。B/S 架构的 Web 应用程序，一般由动态网页负责完成信息呈现和引导交互的任务，JSP 是 Java EE 的体系中编写动态网页的一种技术，而 HTML 则是 JSP 一个不可或缺的组成部分，所以本章先介绍 HTML 的知识(注：这部分工作主要由前端工程师完成，所以本章介绍的内容仅为后文服务，建议感兴趣的同学课后进行扩展阅读)。

2.1 HTML 语法基础

HTML(HyperText Markup Language)翻译过来就是超文本标记语言，它是为“网页创建和其他可在网页浏览器中看到的信息”设计的一种标记语言。目前的版本为 HTML 5。

HTML 文档也被称为网页(webpage)，保存 HTML 文档时，文档后缀可使用.htm，也可使用.html。它编写比较简单，整个文档由 HTML 元素构成，可以说，掌握了 HTML 元素的写法，就掌握了 HTML 的全部。

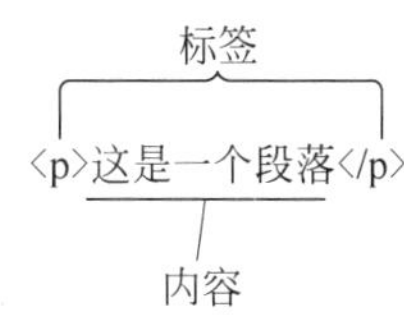

图 2.1 典型的 HTML 元素示意图

1. HTML 元素

先看一个典型的 HTML 元素的例子，如图 2.1 所示。

```
<p>这是一个段落</p>
```

这个元素由两部分构成：标签、内容。

<p></p>这一组叫作标签，<p>称为开始标签，</p>称为结束标签，略一观察便知两者之间的关联，差别就在于结束标签的字母p前多了一个左斜杠/。

文字部分“这是一个段落”属于内容。

2. HTML 标签

标签通常都是成对出现，但是也有例外，当元素内容为空时，称为空元素。空元素在开始标签中进行关闭(以开始标签的结束而结束)，例如
、<hr />。

> 注1：标签的闭合
> 在实际编程中，我们能看到有些代码是
、<hr />的，也能正确运行，展示效果无差别，这是因为我们见到的代码有些遵循的是HTML的语法标准，有的则遵循的是XHTML的标准，在XHTML中这些标签必须是闭合的。建议遵循XHTML的语法规则。(XHTML是HTML与XML结合的产物)
> 注2：XHTML与HTML的主要区别
> XHTML元素必须被正确地嵌套。
> XHTML元素必须被关闭。
> 标签名必须用小写字母。
> XHTML文档必须拥有根元素。
> 属性名称必须小写。
> 属性值必须加引号。
> 属性不能简写。
> 注3：关于HTML、XHTML、HTML5三者的介绍及比较，可以参考http://www.w3school.com.cn/h.asp上的介绍。
> 注4：NetBeans IDE中既可以创建HTML文档，也可以创建XHTML文档，创建文档类型不同，语法检查时遵循的标准也不同。

HTML标签都是有实际意义的，例如：<p></p>表示段落，取自段落一词的英文paragraph缩写；<hr />显示为水平线，取自水平线一词horizontal的缩写。在学习的过程中要注意总结规律，避免死记硬背。

3. HTML 属性

HTML标签可以拥有属性。属性提供了有关HTML元素的更多信息。属性总是以名称/值成对的形式出现，例如：name = "value"。属性总是在HTML元素的开始标签中规定。我们为上一个例子中的<p></p>标签添加一个对齐属性，让这段文字显示方式为左对齐。代码如下：

```
<p align = "left">这是一个段落</p>
```

属性名为align，属性值为left，属性值取值不唯一，也可以是right、center、justify。在当前企业级的开发中，除基础属性外，其他与格式控制及形状显示相关的属性，都不建议直接赋值，而建议使用CSS。CSS的学习是另一个话题，本书不详述。

4. HTML 文档

HTML文档的组织有一个通用的结构，如图2.2所示。

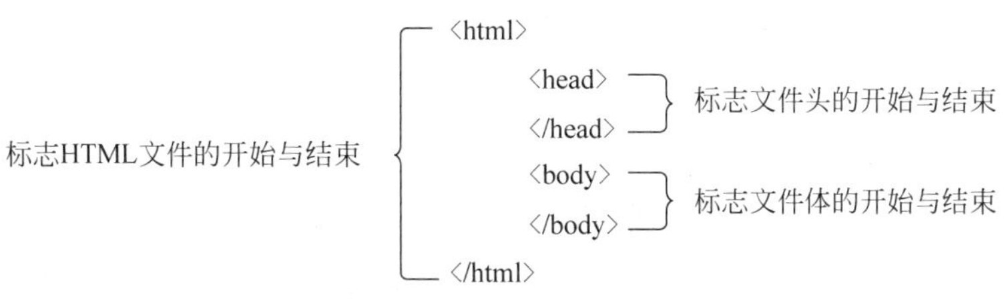

图 2.2 HTML 文档通用结构

在 HTML 格式的文件中，有些元素只出现在文件头中，而大多数元素都出现在文件体中。头部信息一般不显示，可供搜索引擎抓取，或用来对文档的一般信息进行描述。文件体中的信息则多数会经由浏览器解析显示给用户。

HTML 元素可以分成两大类：一大类侧重于信息呈现，作用是把需要展示的信息展示给用户，偏静态展示；另一大类侧重于引导用户提交信息，然后再由服务器根据用户不同，做出不同的处理，实现动态的效果。

2.2 使用 NetBeans IDE 创建 Java Web 应用程序

创建一个 Java Web 应用程序来完成 HTML 部分内容的学习和练习。

目标：创建一个 Java Web 应用程序，完成 HTML 部分内容的学习和练习。

工程名：eg0201。

打开 NetBeans IDE，在菜单栏中选择“文件”→“新建项目”命令，在“新建项目”对话框的“类别”列表中选择 Java Web 选项并在右边的“项目”列表中选择“Web 应用程序”项目，如图 2.3 所示。

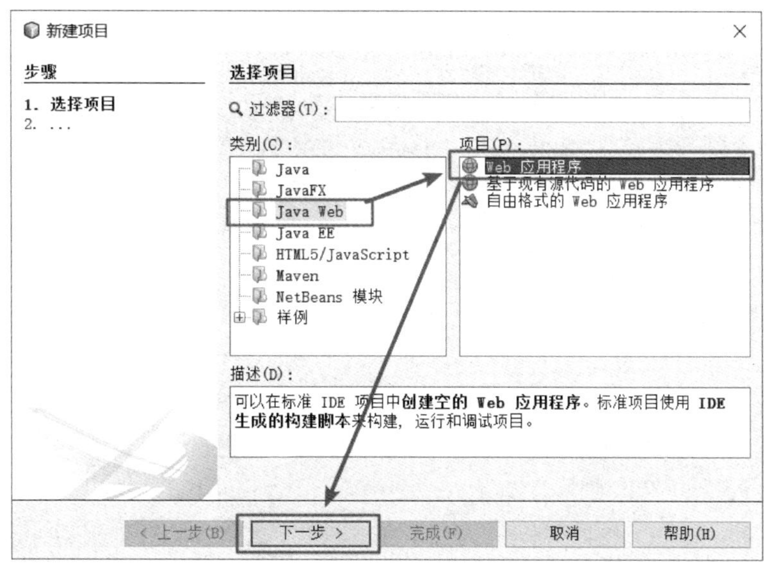

图 2.3 新建 Web 应用程序

设置项目名称和存放位置，这里项目名称为 eg0201，位置为 C:\NetBeansProjects，单击“下一步”按钮，如图 2.4 所示。

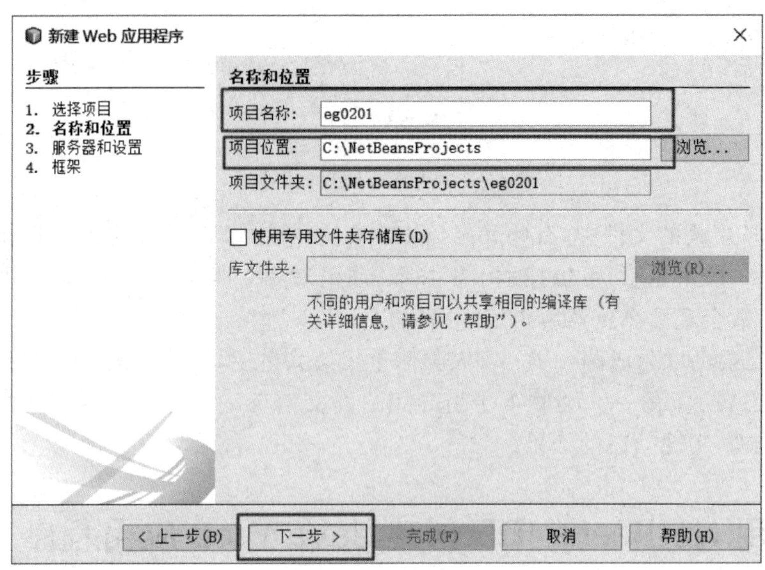

图 2.4 设置项目名称和位置

设置服务器和上下文路径，此处取默认值(课后可以尝试更改上下文路径，观察项目运行时上下文路径到底是什么)。单击“完成”按钮，至此新建 Web 应用程序完成，如图 2.5 所示。

创建完成后在 Web 页目录下会自动创建一个 index.html 的网页文件。如图 2.6 所示。此时可以运行该网页，运行方式与运行主类相同。

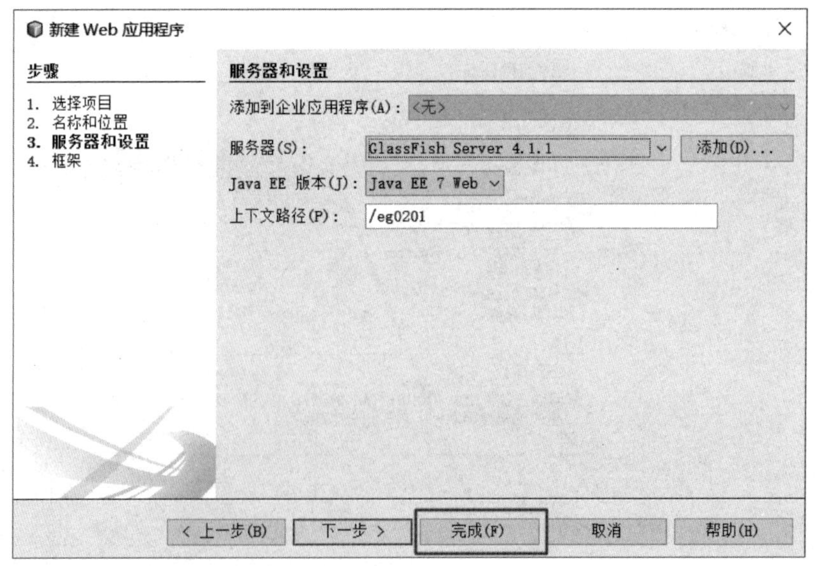

图 2.5 设置服务器和上下文路径

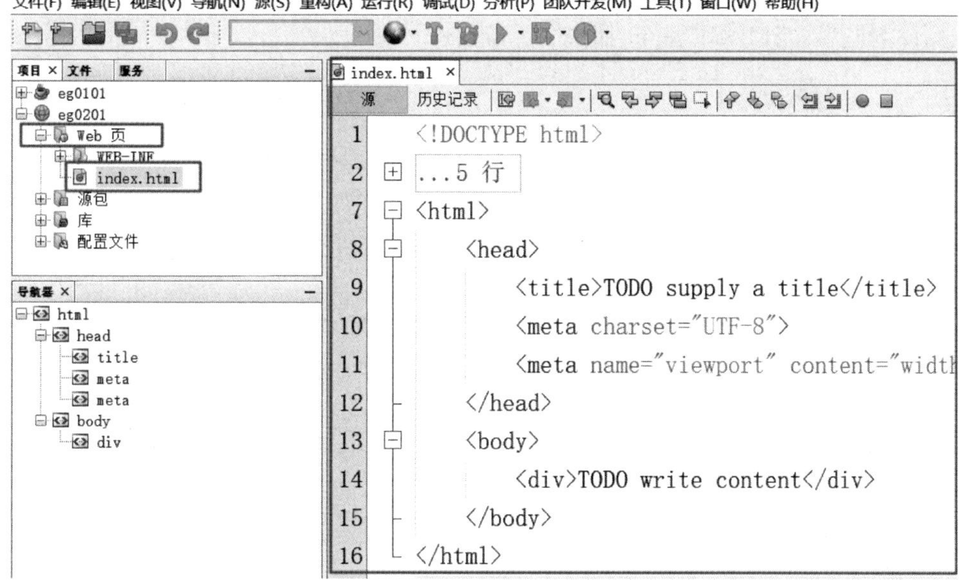

图 2.6 创建 index.html 文件

运行效果如图 2.7 所示。

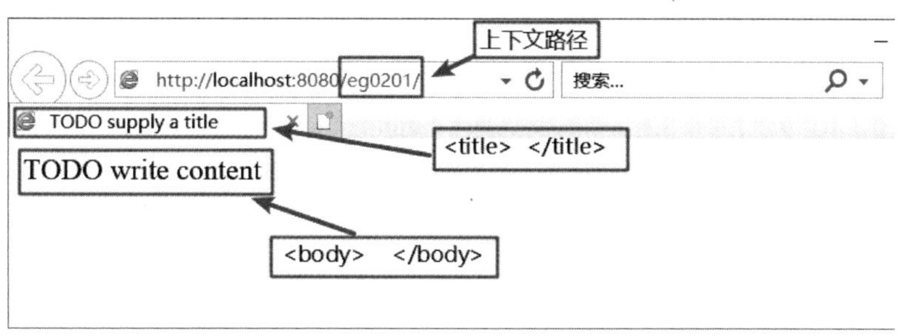

图 2.7 eg0201 中 index.html 运行效果

2.3 侧重信息呈现的 HTML 元素

侧重信息呈现的 HTML 元素主要有文本控制元素、表格相关元素和超级链接元素等。

1. 文本控制元素

1) <p></p>(段落)

表示 paragraph,作用是创建一个段落。

2)
(折行)

表示 line break,作用是折行/强制换行。一个段落标记等价于两个折行。

3) <hr/>(水平线)

表示 horizontal,作用是创建一条水平线,通常用于将页面正文与页脚的版权页分隔开。

在 eg0201 项目中新建一个名为 text、后缀为 html 的文件。创建过程如下：右击“Web 页”，选择“新建”→“HTML”命令，如图 2.8 所示。

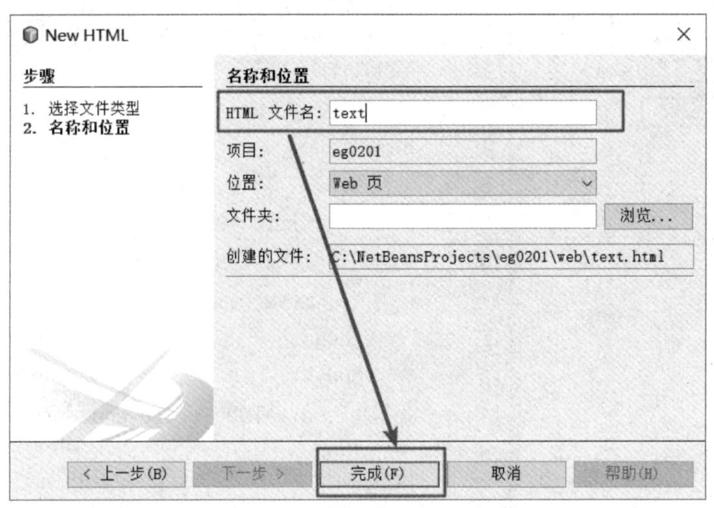

图 2.8　新建 HTML 文档

编写如下代码，并设想一下运行时可能的样子，看看运行效果如何(注：如果想要以单击“运行”按钮方式运行此文档，会发现无法达到目的，运行显示的是 index.html 的内容。因为单击“运行”按钮，实际运行的是项目默认的首页/欢迎页面)。

text.html

```
<!DOCTYPE html>
<html>
    <head>
        <title>文本格式化示例</title>
        <meta charset="UTF-8">
        <meta name="viewport" content="width=device-width, initial-scale=1.0">
    </head>
    <body>
        <p align="left">左对齐的一段</p>
        <p align="center">居中对齐的一段</p>
        <p align="right">右对齐的一段</p>
        <p>这一段无奈地<br/>折行了</p>
        <hr/>
    </body>
</html>
```

2. 表格相关元素

表格通常用作数据显示时的格式化，包括一组标签。所有浏览器都支持的标签有如下几组。

1) <table></table>(表格)

其他的表格标签要嵌套在其中才有效。有两个常用的属性：align(对齐，取值为 left、

center、right)、border(设置表格边框的大小,取值为数字,如 border="1")。

2) <caption></caption>(表格标题)

<caption>标签必须紧随<table>标签之后。只能对每个表格定义一个标题。通常这个标题会居中于表格之上。

3) <tr></tr>(表格的行)

记忆时可记作 table row 的首字母缩写。每个表格有若干行,所以需要嵌套在<table></table>标签内部。

4) <td></td>(表格数据)

记忆时可记作 table data 的首字母缩写。<td></td>元素的内容可以包含文本、图片、列表、段落、表单、水平线、表格等。每行被分割为若干单元格,所以需要嵌套在<tr></tr>标签内部。

5) <th></th>(表格的表头)

记忆时可以记作 table heading。需要嵌套在<tr></tr>标签内部,与<td></td>类似,区别在于表头默认显示形态为加黑居中,以与单元格内容区分开。

下面来看一组示例,通过编写、运行示例体会各个标签的用法。在 eg0201 项目中分别新建 table1.html、table2.html、table3.html、table4.html 四个 HTML 文档,并编写如下代码。

table1.html

```
<!DOCTYPE html>
<html>
    <head>
        <title>无边框的表格</title>
        <meta charset="UTF-8">
        <meta name="viewport" content="width=device-width, initial-scale=1.0">
    </head>
    <body>
        <table>
            <caption>无边框的 2 行 2 列表格</caption>
            <tr>
                <td>第一行,第一个单元格</td>
                <td>第一行,第二个单元格</td>
            </tr>
            <tr>
                <td>第二行,第一个单元格</td>
                <td>第二行,第二个单元格</td>
            </tr>
        </table>
    </body>
</html>
```

table2.html

```
<!DOCTYPE html>
<html>
```

```
    <head>
        <title>有边框的表格</title>
        <meta charset = "UTF - 8">
        <meta name = "viewport" content = "width = device - width, initial - scale = 1.0">
    </head>
    <body>
        <table border = "1">
            <caption>无边框的 2 行 2 列表格</caption>
            <tr>
                <td>第一行,第一个单元格</td>
                <td>第一行,第二个单元格</td>
            </tr>
            <tr>
                <td>第二行,第一个单元格</td>
                <td>第二行,第二个单元格</td>
            </tr>
        </table>
    </body>
</html>
```

table3.html

```
<!DOCTYPE html>
<html>
    <head>
        <title>有表头的表格</title>
        <meta charset = "UTF - 8">
        <meta name = "viewport" content = "width = device - width, initial - scale = 1.0">
    </head>
    <body>
        <table border = "1">
            <caption>有边框的 3 行 2 列表格</caption>
            <tr>
                <th>列 1</th>
                <th>列 2</th>
            </tr>
            <tr>
                <td>第一行,第一个单元格</td>
                <td>第一行,第二个单元格</td>
            </tr>
            <tr>
                <td>第二行,第一个单元格</td>
                <td>第二行,第二个单元格</td>
            </tr>
        </table>
    </body>
</html>
```

table4.html

```
<!DOCTYPE html>
<html>
    <head>
        <title>居中对齐的表格</title>
        <meta charset = "UTF - 8">
        <meta name = "viewport" content = "width = device - width, initial - scale = 1.0">
    </head>
    <body>
        <table border = "1" align = "center">
            <caption>有边框的 3 行 2 列表格</caption>
            <tr>
                <th>列 1 </th>
                <th>列 2 </th>
            </tr>
            <tr>
                <td>第一行,第一个单元格</td>
                <td>第一行,第二个单元格</td>
            </tr>
            <tr>
                <td>第二行,第一个单元格</td>
                <td>第二行,第二个单元格</td>
            </tr>
        </table>
    </body>
</html>
```

3. 超级链接元素

HTML 使用超级链接与网络上的另一个文档相连，单击链接可以从一个页面跳转到另一个页面，这是 HTML 文档与传统文本文档的区别之一。目前，网站的开发者常利用这一特性进行站点的导航。

```
<a href = "url"></a>
```

单纯的<a></a>是没有意义的，需要属性的配合。href 是必不可少的属性。所以一般形式是这样的：

```
<a href = "url">显示在页面上的文字</a>
```

href 属性规定链接的目标。开始标签和结束标签之间的文字被作为超级链接来显示(默认是下画线状态，鼠标移动在其上时，变成手形)。下面通过一个例子学习一下超级链接的基本用法。

link1. html

```
<!DOCTYPE html>
<html>
    <head>
        <title>超级链接的练习</title>
        <meta charset="UTF-8">
        <meta name="viewport" content="width=device-width, initial-scale=1.0">
    </head>
    <body>
        <a href="https://www.baidu.com">点此到达百度首页</a>
        <a href="index.html">单击到达首页 index.html</a>
    </body>
</html>
```

上面这行代码运行结果如图 2.9 所示。

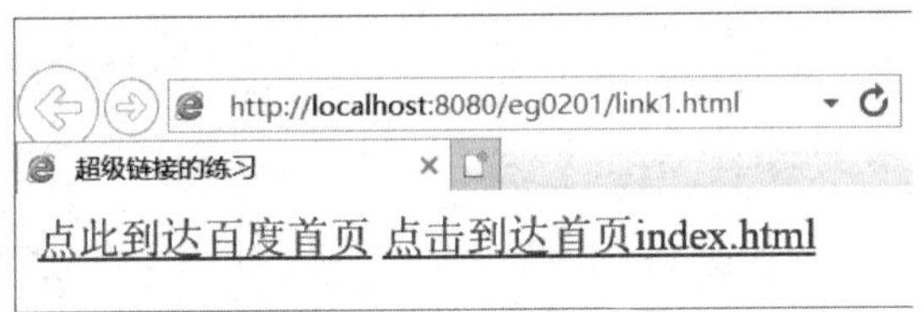

图 2.9　超级链接练习结果

单击第一个链接在联网状态会访问百度首页，单击第二个链接会显示 index.html 的内容。

在上面的例子中，链接会在当前窗口打开，经常上网的话一定会有这样的经历，单击一个链接，链接的目标在当前窗口中打开，或者在浏览器新的标签页打开了。控制打开方式的属性是 target 属性。使用 target 属性，你可以定义链接指向的网页在何处显示，属性值可以是_blank、_parent、_self、_top、某个框架的名字(框架不在本课程讨论范围内，感兴趣的读者可自学)，最常用的是_blank。

新建一个 HTML 文档 link2.html，在 link1.html 代码基础上添加 target 属性，看看运行结果有什么变化？

link2. html

```
<!DOCTYPE html>
<html>
    <head>
        <title>超级链接的练习,target 属性</title>
        <meta charset="UTF-8">
        <meta name="viewport" content="width=device-width, initial-scale=1.0">
    </head>
    <body>
        <a href="http://www.baidu.com" target="_blank">
            点此到达百度首页
```

```
        </a>
        <a href = "index.html" target = "_blank">单击到达首页 index.html </a>
    </body>
</html>
```

注意：<a href="url">显示在页面上的文字</a>中，“显示在页面上的文字”不必一定是文本。图片也可以成为链接(例子略)。

2.4 侧重引导用户提交信息的HTML元素

HTML表单元素用来引导用户提交信息。可以将它视为平时纸质表格的电子化。

1. <form></form>元素

<form></form>通常翻译成表单或者窗体，是其他表单元素的根元素。是用来告诉浏览器，这组标签之间的内容是要提交的，其他的表单元素要嵌套在这组标签之间才能被提交。一个HTML文档中可以有多个表单，单击某个提交按钮，则该提交按钮所在的表单内容会被提交到服务器。

当用户单击提交按钮时，表单的内容会被传递到另一个组件。传递给谁由<form>标签的动作属性(action)决定，取值为目的组件的url-pattern(中文译为访问模式，对于JSP页面而言，通常是含后缀的文件名)。由动作属性定义的这个组件会对接收到的输入数据进行相关的处理。例如，已知提交的信息要由同文件夹内的handle.jsp处理，那么<form>标签的action属性应该这样设置：

```
<form action = "handle.jsp">
    代码略
</form>
```

如果要由mytest子文件夹中的handle.jsp处理，那么<form>标签的action属性应该先写子文件夹名，再写文件名，就像这样：

```
<form action = "mytest/handle.jsp">
    代码略
</form>
```

除了动作属性action，<form>标签还有一个很重要的属性，那就是设置信息提交方式的属性method。method的取值固定，有两个选择(get或者post)。如果该属性不设置的话，默认为get(建议在编程过程中明确指定属性值为post)。两者区别主要在于：

(1) 传送的数据量大小不一样。get有限制；post理论上不限制大小。

(2) 安全性不同。get提交的数据会在浏览器地址栏以明文显示出来，容易暴露；post不以明文形式提交，所以更安全。

典型的< form ></form >元素写法可能是这样的：

```
<form action = "handle.jsp" method = "post">
    代码略
</form>
```

注意：除了在表单中明确设置了 method="post"外，无论是单击超级链接，还是在浏览器地址栏直接输入地址，其请求的提交方式都为 get，请注意分辨。

2. < input />元素

< input />元素用来引导用户输入信息，它是一个空元素，所以最好写成闭合的形式。可以通过设置< input />标签属性 type 的值，来控制它在页面显示出来的形态。

一个< form >内可以有多个< input />元素，为了区分各个元素，每个元素有一个名字属性 name 需要设置。此外，有些元素需要赋初值，这时候还需要设置值属性 value。典型的< input />元素写法如下：

```
<input type = "类别" name = "名字" value = "初值" />
```

其中 type 属性取值集合固定，可以是 text、password、radio、checkbox、file、submit、reset、button、image、hidden，下面分别介绍：

1) text(文本字段)

```
<input type = "text" name = "textfield" value = "请在此输入" />
```

2) password(密码字段)

与文本字段类似，但不显示输入的内容，以一串●显示。

```
<input type = "password" name = "mypassword" />
```

3) radio(单选按钮)

单选按钮比较特殊，尽管一般在页面上显示多个，但是代表的是同一个控件，所以命名时 name 属性应该取相同的值，同一时刻只能选中其中一个，例如：性别的选择，或者单项选择题的选项。

```
<input type = "radio" name = "gender" />
<input type = "radio" name = "gender" />
```

上面的代码看起来肯定有些怪异，简直是一模一样，实际应用中当然不是这样。每一个呈现出来的按钮都应该有不同的值，例如“男”“女”，所以应该对 value 属性进行设置，同时在页面显示的文字中给予提示。为了比较其中的差异，下面例子中刻意设置 value 属性的值与页面显示的值不同，希望大家能注意。网页上显示出来的是 male 和 female，而提交的数据对应的是“男”“女”。

```
<input type="radio" name="gender" value="男"/>male
<input type="radio" name="gender" value="女"/>female
```

此外，如果想要页面打开时就预先选中某个值，可以设置 checked 属性，一般设置为：

```
<input type="radio" name="gender" value="男" checked="checked" />male
```

HTML 的标准设置为：

```
<input type="radio" name="gender" value="男" checked />male
```

4) checkbox(复选框)

复选框与单选按钮类似，只是应用领域为多选的情况(可以选择 0 至多个值)，例如下面例子中水果的种类，就使用了 checkbox。同一组 checkbox 的 name 属性值一般设置成同一个值，视为同一控件，示例如下：

```
<input type="checkbox" name="fruits" value="苹果"/>苹果
<input type="checkbox" name="fruits" value="香蕉"/>香蕉
<input type="checkbox" name="fruits" value="橘子"/>橘子
```

虽然也可以设置成下面代码的形式，视为不同的控件，但是就失去了复选框的本意，所以在使用中，建议同一组 checkbox 的 name 属性值设置成相同的值。

```
<input type="checkbox" name="apple" value="苹果"/>苹果
<input type="checkbox" name="banana" value="香蕉"/>香蕉
<input type="checkbox" name="orange" value="橘子"/>橘子
```

如需默认选中某一项或者某几项，同样设置 checked 属性，与单选按钮一样，此处不赘述。

5) file(文本域)

文本域主要用于文件上传。

```
<input type="file" name="attachfile" />
```

6) submit(提交按钮)

提交按钮会做出提交数据的动作。一般不需要取得该按钮的值，所以 name 属性通常不设定。而为了给用户明确的提示，需要设置 value 属性，该属性的值会显示在按钮上。例如，最普遍的状况是这样：

```
<input type="submit" value="提交" />
```

7）reset（重置按钮）

重置按钮会将已填写的表单数据清空。与提交按钮类似，一般不需要取得该按钮的值，所以 name 属性通常也不设定。为了给用户明确的提示，需要设置 value 属性，该属性的值会显示在按钮上。例如，最普遍的状况是这样：

```
<input type="reset" value="重置" />
```

8）button（普通按钮）

普通按钮需要结合脚本才能完成提交的动作，其他与提交按钮、重置按钮类似，初学不建议使用。

```
<input type="button" value="一个按钮" />
```

9）image（图片按钮）

图片按钮功能与提交按钮类似，只是按钮呈现出来的不是 value 属性设置的值，而是 src 属性设置的图片。可以通过设置 alt 属性来设置图片失效时显示的文字。

```
<input type="image" src="bt1.png" alt="图片按钮" />
```

10）hidden（隐藏域）

隐藏域作用与文本字段类似，只是不显示出来，用于悄悄传递数据，通常需要设置 value 属性。

```
<input type="hidden" name="hiddenfield" value="somevalue"/>
```

在 NetBeans IDE 中创建一个名为 form1.html 的文档。运行并注意观察代码中属性的设置与网页上控件形态的对应关系。

form1.html

```
<!DOCTYPE html>
<html>
    <head>
        <title>输入元素 input</title>
        <meta charset="UTF-8">
        <meta name="viewport" content="width=device-width, initial-scale=1.0">
    </head>
    <body>
        <form action="handle.jsp" method="post">
        用户名<input type="text" name="username" /><br />
        密码<input type="password" name="userpass" /><br />
        性别<input type="radio" name="gender" value="男" checked="checked" />男
            <input type="radio" name="gender" value="女" />女<br />
```

```
        爱好<input type="checkbox" name="hobby" value="reading" />读书
            <input type="checkbox" name="hobby" value="sports" checked="checked" />体育
            <input type="checkbox" name="hobby" value="swimming" />游泳
            <br />
        上传头像<input type="file" name="uploadfile" /><br />
        提交按钮<input type="submit" value="提交" /><br />
        重置按钮<input type="reset" value="重置" /><br />
        按钮<input type="button" value="你点我呀" onclick="" /><br />
        图片按钮<input type="image" src="" alt="图片按钮" /><br />
        隐藏域<input type="hidden" name="hiddenfield" />
        </form>
    </body>
</html>
```

还有两个常用的表单元素，用来收集用户的信息，一个是文本区域，一个是列表框，但它们的标签不是<input />。

3. <textarea></textarea>元素

<textarea>标签定义的是多行文本输入控件，也称文本域或者文本区域。文本域可容纳更多的文本，适用于大量文本的输入。可以通过 cols 和 rows 属性来控制 textarea 的外观，不过更好的办法是使用 CSS 来设置。

在 NetBeans IDE 中创建一个名为 form2.html 的文档。运行并注意观察代码中属性的设置与网页上控件形态的对应关系。尝试在页面的文本区域中输入更多的文字，注意观察运行结果。

form2.html

```
<!DOCTYPE html>
<html>
    <head>
        <title>文本区域 textarea</title>
        <meta charset="UTF-8">
        <meta name="viewport" content="width=device-width, initial-scale=1.0">
    </head>
    <body>
        <form action="handle.jsp" method="post">
            <textarea name ="introduction" rows="2" cols="15">
                这个人很懒,什么也没有留下
            </textarea>
        </form>
    </body>
</html>
```

4. <select></select>元素

<select></select>元素可以创建单选或多选的下拉列表。典型用法如下例所示：

form3. html

```
<!DOCTYPE html>
<html>
    <head>
        <title>下拉列表 select </title>
        <meta charset = "UTF-8">
        <meta name = "viewport" content = "width = device - width, initial - scale = 1.0">
    </head>
    <body>
        <form action = "handle.jsp" method = "post">
            <select name = "birth_place" size = "1">
                <option value  = "H">黑龙江</option>
                <option value  = "J">吉林</option>
                <option value = "L">辽宁</option>
            </select>
        </form>
    </body>
</html>
```

其中<option></option>元素为下拉选项，如果想要默认选中某个选项，则需要设置 selected。例如，上例中如果要默认选中辽宁，则应这样设置：

```
<option value = "L" selected = "selected">辽宁</option>
```

2.5 本章回顾

本章先介绍了 HTML 的语法基础知识，如图 2.10 所示

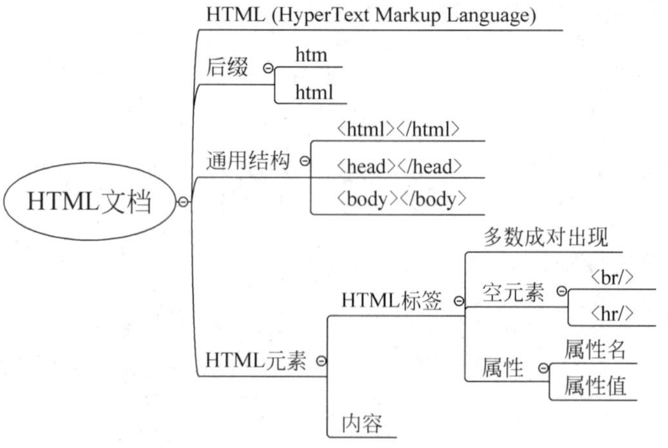

图 2.10 HTML 语法基础知识

接着介绍了与信息呈现相关的 HTML 元素，其中包括文本控制元素、超级链接元素、表格相关元素，如图 2.11 所示。

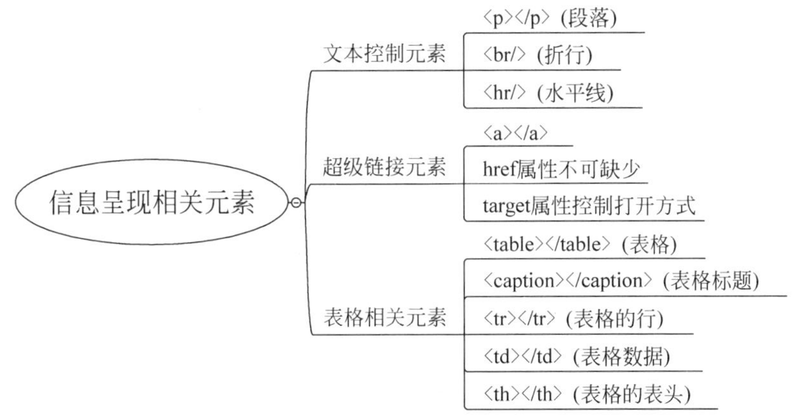

图 2.11 HTML 信息呈现相关元素

最后介绍了侧重引导用户提交信息的 HTML 元素——表单元素，如图 2.12 所示。

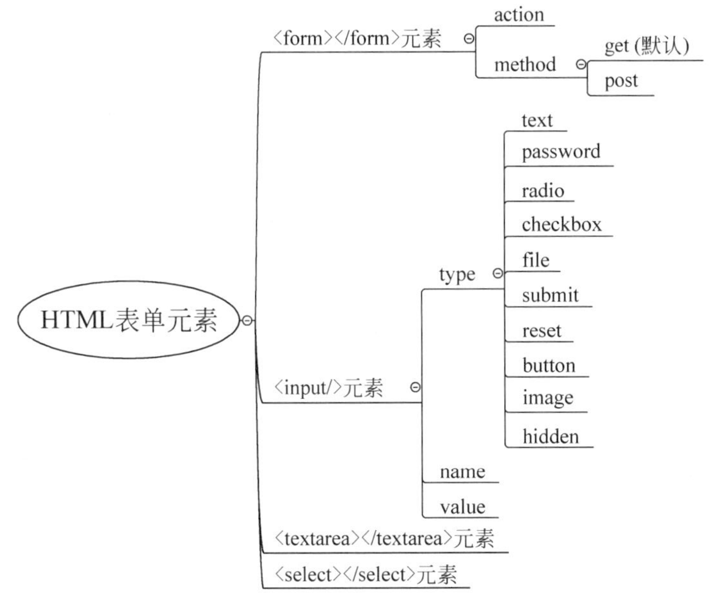

图 2.12 HTML 表单元素

2.6 课后习题

1. 名词解释：HTML。
2. HTML 元素由几部分构成，分别是什么？
3. 标签都是成对出现的吗？如果不是，举出几个反例。
4. 段落标签< p />与折行标签< br />有什么区别和联系？

5. 与表格相关的标签有哪些？试编写一下你的课程表。

6. 试用超级链接编写一个网站的首页导航。

7. 试用表单元素完成一个注册页面，假设交由 reg_handle.jsp 处理，要求有一定的安全性。

8. 以 get 方式提交数据与以 post 方式提交数据有什么区别？

9. 上下文路径对应浏览器地址栏 URL 中的哪个部分？

第3章 Servlet编程基础

学习目标：

通过本章的学习，你应该：

- 了解 HTTP 通信协议基本概念
- 了解 Servlet 的定义、功能和用途
- 掌握如何通过 NetBeans IDE 进行 Servlet 编程
- 掌握 Servlet 处理 HTTP 请求的流程
- 掌握设置 HTTP 响应头信息方法
- 掌握获取 HTTP 请求报头方法
- 掌握请求对象获取用户通过超级链接或表单提交的信息的方法
- 掌握在 Servlet 中处理表单通过 post 方式提交的信息中“中文乱码”问题
- 掌握响应对象对响应进行重定向的方法
- 掌握请求转发器转发请求的两个方法
- 理解 Servlet 的生命周期
- 了解 Servlet 的部署方式

一点提示：Servlet 的相关内容是非常重要的，掌握了 Servlet，学习 JSP 也会事半功倍，所以希望大家给予足够的重视。在本章中，并不深入探讨 Servlet 的理论知识，而是从应用入手，先通过 Servlet 技术完成一些简单而实用的操作，程序调试成功带来的小得意总好过一头雾水硬着头皮啃下去。唯一的要求就是：跟着书上的例子做一下。

Web 应用程序是基于 HTTP/HTTPS 的，所以先来了解一下 HTTP/HTTPS。

3.1 HTTP/HTTPS 通信协议基本概念

上网的时候，一般从用户的角度来说，其实并不关心某个网站是用什么技术实现的，采用了什么架构，大家更在乎的是网站的内容。用户操作的流程一般是这样的：打开浏览器，输入网址（或者通过某种快捷方式，例如：打开收藏夹），然后就开始了网上冲浪的旅程。一切都是那么的简单，例如：想访问百度，就输入 www.baidu.com ，想访问淘宝就输入 www.taobao.com 。但是实际操作时，注意观察一下就会发现，浏览器在这个过程中帮你做了一个补足的操作，地址栏里面真正显示出来的是 https://www.baidu.com/和 https://www.

taobao.com/ 。有共性没？有，前面多加了一个“https://”。(其实尾部还多了一个/，起什么作用请自行搜索 URL 格式)。

HTTPS(Hyper Text Transfer Protocol over Secure Socket Layer)是以安全为目标的 HTTP 通道，简单讲是 HTTP 的安全版。

HTTP(Hyper Text Transfer Protocol 超文本传输协议)是互联网上应用最为广泛的一种网络协议。所有的 WWW 文件都必须遵守这个标准。

设计 HTTP 最初的目的是为了提供一种发布和接收 HTML 页面的方法，浏览器就是通过这一协议与服务器交互。

HTTP 允许用户提出 HTTP“请求”(HTTP request)，然后由服务器根据实际处理的结果传回 HTTP“响应”(HTTP response)，它的基本运行方式为：

(1) 当用户向 Web 服务器发送请求时，Web 服务器将会开启一个新的连接，通过这个连接，用户可以将 HTTP 请求传送给 Web 服务器。

(2) 当 Web 服务器收到 HTTP 请求时，将进行解析与处理，并将处理结果包装成 HTTP 响应。

(3) 最后，Web 服务器会将 HTTP 响应传送给用户，只要用户接收到 HTTP 响应，Web 服务器就会关闭这个连接，用户的执行状态将不会被保存。

这个过程可以简化成“请求—处理—响应”的模型，简化模型如图 3.1 所示。本章学习的 Servlet 就是 Web 服务器端负责处理请求，并产生响应的一个组件。

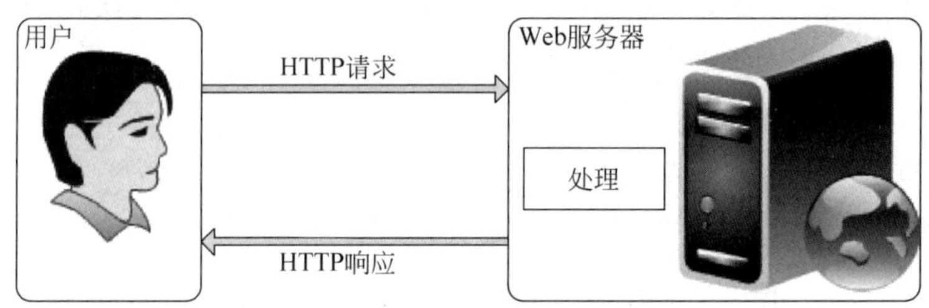

图 3.1 HTTP 运行方式简化模型“请求—处理—响应”

3.2 Servlet 的定义及作用

Servlet(Server Applet)是 Java Servlet 的简称，称为小服务程序或服务连接器，用 Java 编写的服务器端程序，主要功能在于交互式地浏览和修改数据，生成动态 Web 内容。

Servlet 的主要作用如下：

(1) 通过请求对象读取用户程序发送过来的显式数据(如表单数据)。

(2) 通过请求对象读取用户程序发送过来的隐式数据(如请求报头)。

(3) 处理数据并生成响应内容或设置响应报头。

在 Servlet 中，请求对象的类型为 HttpServletRequest，响应对象的类型为 HttpServletResponse。

3.3 使用 Servlet 生成一个网页

目标：创建一个 Java Web 应用程序，使用 Servlet 生成一个网页。

工程名：eg0301。

打开 NetBeans IDE，新建一个 Web 应用程序，具体过程可参考 1.2 节相关内容。在创建好的 eg0301 工程里，右击“源包”，在弹出的菜单中选择“新建→Servlet”命令，如图 3.2 所示。

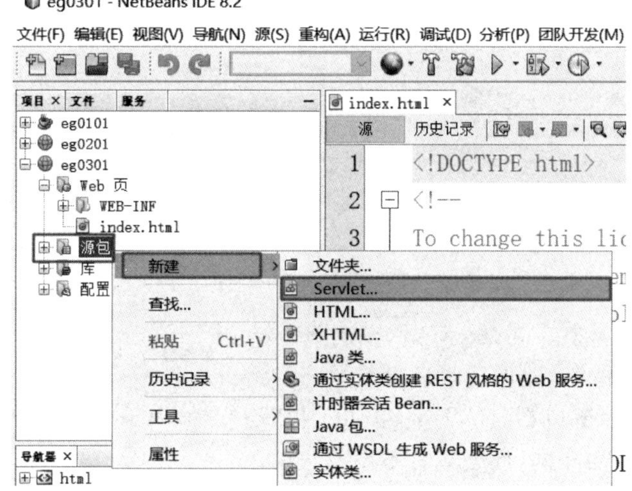

图 3.2 新建一个 Servlet

然后设置 Servlet 的名称和位置，此处类名暂时保持不变，包名设置为 cn.edu.djtu，然后单击“完成”按钮，如图 3.3 所示。

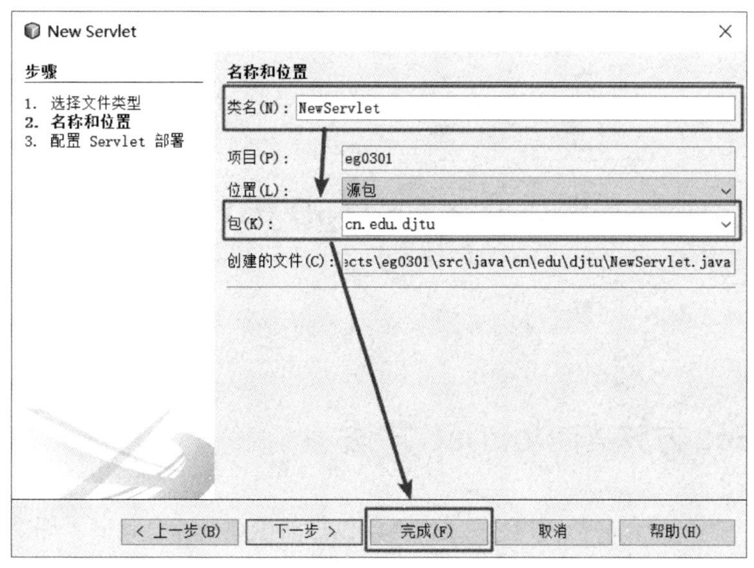

图 3.3 设置 Servlet 的名称和位置

完成后的效果如图 3.4 所示。

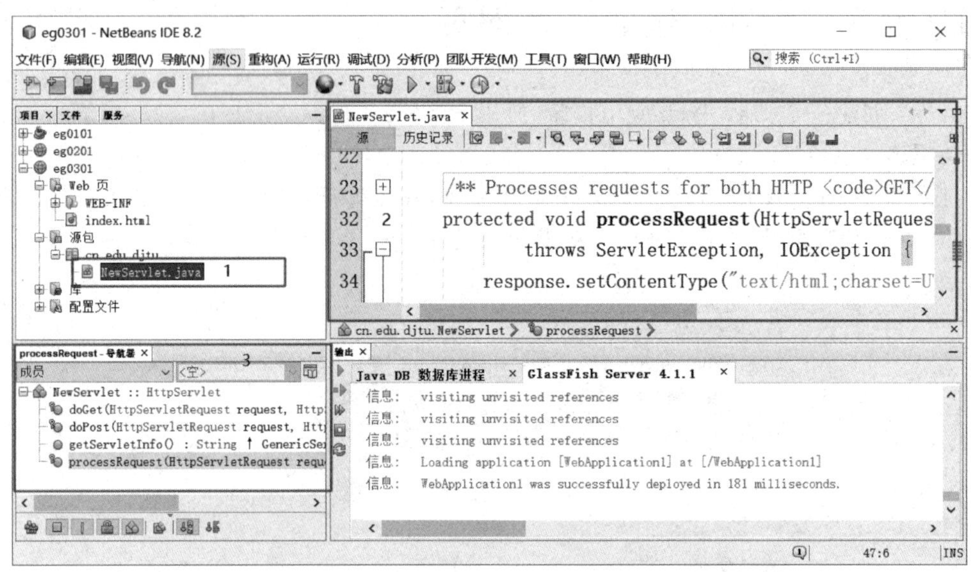

图 3.4　创建完 Servlet 后的状态

在图 3.4 区域 1 中右击 NewServlet.java 文档，选择“运行文件”命令，即可观察运行效果，运行会弹出如图 3.5 所示窗口，提示设置 Servlet 执行 URI，默认与 Servlet 名相同，此处暂时不设置，直接单击“确定”按钮。

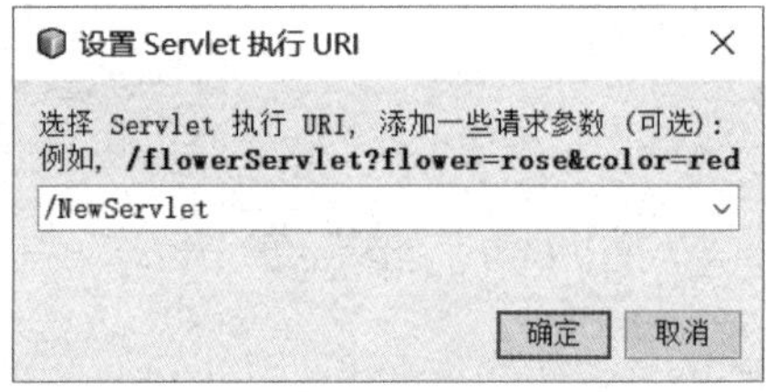

图 3.5　设置 Servlet 执行 URI

运行效果如图 3.6 所示。

至此，就创建了一个名为 NewServlet.java 的 Servlet 文件，它的 urlPattern 是 NewServlet。当有用户请求该 Servlet 时，它生成一个网页返回给用户。

图 3.6　NewServlet 运行效果

3.4　doGet 方法与 doPost 方法

在 eg0301 中，NetBeans IDE 会根据代码模板自动生成一部分代码。图 3.4 所示区域 3 中可以直观便利地查看 NewServlet 类的结构。在图 3.4 中可以看到，创建 NewServlet 时已经默认写好了 4 个方法,方法名分别为 doGet、doPost、getServletInfo、processRequest。

其中 getServletInfo 方法可以返回对本 Servlet 的一个简短的描述,通常不需要实现具体业务逻辑,在实际应用中一般不必理会,默认其代码也是折叠的。

doGet 方法响应用户通过 get 方式提交的请求,代码如下:

```
@Override
protected void doGet(HttpServletRequest request, HttpServletResponse response)
    throws ServletException, IOException {
        processRequest(request, response);
    }
```

doPost 方法响应用户通过 post 方式提交的请求,代码如下:

```
@Override
protected void doPost(HttpServletRequest request, HttpServletResponse response)
    throws ServletException, IOException {
        processRequest(request, response);
    }
```

可见在 NetBeans IDE 中,doGet 和 doPost 方法都是调用了 processRequest 方法。所以在 NetBeans IDE 中 doGet 和 doPost 方法代码也是折叠起来的。(其他开发环境未必如此)。

那么除了 get 方式和 post 方式,客户端是否有其他提交信息的方式呢? Servlet 中又是否有对应的 doXxx 方法呢? 答案是:有。

在 NewServlet 源码中的类代码空白处输入 do,然后按下快捷键"Ctrl+\"。这时会发现还有 doDelete、doHead、doTrace、doPut、doOptions 五个方法,如图 3.7 所示。

```
@WebServlet(name = "NewServlet", urlPatterns = {"/NewServlet"})
public class NewServlet extends HttpServlet {
do  ← 输入do, 然后按快捷键" CTRL + \"
doDelete(HttpServletRequest req, HttpServletResponse resp) - 覆盖  void
doHead(HttpServletRequest req, HttpServletResponse resp) - 覆盖  void
doOptions(HttpServletRequest req, HttpServletResponse resp) - 覆盖 void
doPut(HttpServletRequest req, HttpServletResponse resp) - 覆盖  void
doTrace(HttpServletRequest req, HttpServletResponse resp) - 覆盖  void
```

图 3.7 doGet/doPost 以外的 doXxx 方法

这些方法虽然存在,但是很少有机会用到。在实际开发时,基本上只需要覆盖 doGet 或者 doPost 即可。在 NetBeans IDE 中,无论客户端是以 get 方式还是 post 方式提交请求,只要在 processRequest 方法中编写业务逻辑代码就可以了,都可以对客户端的请求做出正确响应。

3.5 使用 Servlet 生成服务器响应

在 eg0301 中,NewServlet 在服务器端生成了一个响应,形式是一个网页。具体代码如下:

NewServlet.java 代码片段

```
protected void processRequest(HttpServletRequest request, HttpServletResponse response)
    throws ServletException, IOException {
    response.setContentType("text/html;charset = UTF - 8");
    try (PrintWriter out = response.getWriter()) {
        /* TODO output your page here. You may use following sample code. */
        out.println("<!DOCTYPE html>");
        out.println("<html>");
        out.println("<head>");
        out.println("<title>Servlet NewServlet</title>");
        out.println("</head>");
        out.println("<body>");
        out.println("<h1>Servlet NewServlet at" + request.getContextPath() + "</h1>");
        out.println("</body>");
        out.println("</html>");
    }
}
```

这段代码是通过代码模板自动生成的,在编写自己的应用时,只要删除 try{}块中的代码,改为自己需要的业务逻辑代码即可。这段代码中,processRequest 方法参数列表中的形参 request 代表 HTTP 请求对象,response 代表 HTTP 响应对象。processRequest 方法体中,代码的第一条语句是:

```
response.setContentType("text/html;charset = UTF - 8");
```

通过响应对象 response 设置了响应的类型为网页形式,同时设置了编码格式为 UTF-8;接下来取得了一个 PrintWriter 类型的对象 out;最后通过 out 对象向客户端输出了一个网页。

除了生成类型为"text/html"形式的响应,也可以生成其他形式的响应,例如 Excel。这时需要设置响应的类型为"application/vnd.ms-excel"。新建一个 Servlet 文件,文件名为 NewServlet1.java。其他参照 NewServlet 设置。运行 NewServlet1,会发现客户端接收到的是 Excel 文档。

NewServlet1.java 代码片段

```
protected void processRequest(HttpServletRequest request, HttpServletResponse response)
    throws ServletException, IOException {
    response.setContentType("application/vnd.ms - excel");
    try (PrintWriter out = response.getWriter()) {
        /* TODO output your page here. You may use following sample code. */
```

```
            out.println("1\t2");
            out.println("3\t4");
        }
    }
```

输出语句中的"\t"起的作用是在 Excel 中将 1 和 2 两个数字分隔在两个单元格里。这个样例不具有实用价值，实际生成 Office 文档时需要考虑的因素很多，所以通常都采用第三方的包，例如 Apache 的 POI 。

"text/html"和"application/vnd. ms-excel"都是 MIME 类型的一员。MIME (Multipurpose Internet Mail Extensions)是描述消息内容类型的因特网标准。MIME 消息能包含文本、图像、音频、视频以及其他应用程序专用的数据。

除了设置内容类型(就是 MIME 类型)，还可以生成其他类型的响应头，在这里就不一一举例了。可以尝试将 NewServlet. java 中增加一行代码。

```
response.sendError(404);
response.setContentType("text/html;charset = UTF - 8");
```

再次运行 NewServlet. Java 观察结果有什么变化。

3.6 使用 Servlet 读取请求报头

尽管在运行 NewServlet. java 时，看起来客户端并没有提交任何信息，其实还是有很多信息被提交了。这部分信息可以被服务器取得，一般不对其进行数据操作，主要就是请求报头的信息。

取得请求报头的信息需要用到请求对象，主要用到它的三个方法，如表 3.1 所示。

表 3.1 javax. servlet. http. HttpServletRequest 定义的方法

方法名	返回值类型	适用情况
getHeader(String name)	String	取得特定请求报头的信息，返回值为字符串
getHeaders(String name)	Enumeration	取得特定请求报头的信息，返回值为枚举类型的对象
getHeaderNames()	Enumeration	取得 HTTP 请求内所有请求报头的信息，返回值为枚举类型的对象

下面通过示例演示如何通过 Servlet 来处理客户端提交的请求报头信息。

目标：学习使用 Servlet，读取客户端提交的请求报头信息。

工程名：eg0302。

用到的文件如表 3.2 所示。

表 3.2 eg0302 用到的文件及文件说明

文件名	说明
ShowHeadServlet. java	手动创建的 Servlet，用来取得用户提交的请求报头信息，并以网页形式显示

编程思路：先取得全部的请求报头名，然后通过请求报头名取得对应的请求报头的值并输出。

打开 NetBeans IDE，新建一个名为 eg0302 的 Java Web 项目，具体过程略；新建名为 ShowHeadServlet.java 的 Servlet，具体过程略。编写各部分代码如下(仅列出核心代码，大部分自动生成的代码略)：

ShowHeadServlet.java 代码片段

```
protected void processRequest(HttpServletRequest request, HttpServletResponse response)
    throws ServletException, IOException {
    response.setContentType("text/html;charset = UTF - 8");
        try (PrintWriter out = response.getWriter()) {
        Enumeration < String > allHeaderNames = request.getHeaderNames();
            while (allHeaderNames.hasMoreElements()) {
                String headerName = allHeaderNames.nextElement();
                String headerValue = request.getHeader(headerName);
                out.print(headerName + " : " + headerValue);
                out.print("< hr />");
            }
        }
    }
```

运行 ShowHeadServlet，运行结果如图 3.8 所示。

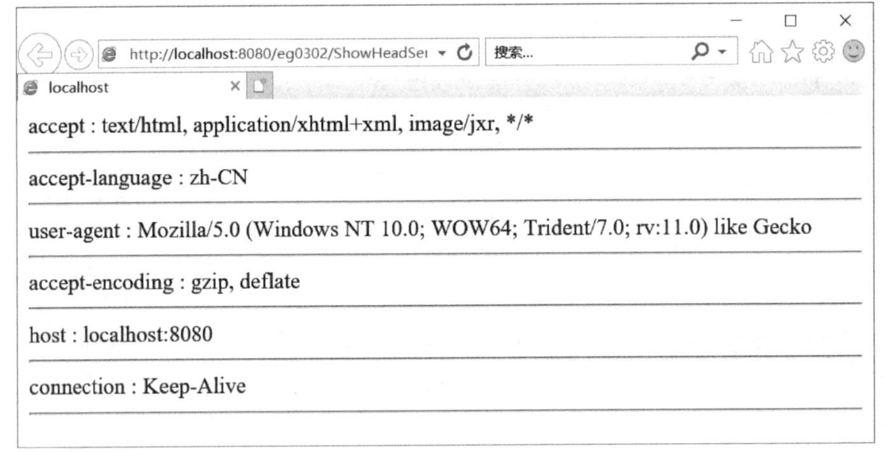

图 3.8 ShowHeadServlet 运行结果

3.7 使用 Servlet 读取用户通过超级链接传送的信息

在学习之前先看一个例子。打开百度首页，在右上角选择“更多产品”，里面有“糯米”，单击“糯米”即可到达“百度糯米”首页。观察地址栏可以看到 URL 中有“? cid=002540”，如图 3.9 所示。说明从百度首页跳转到百度糯米首页时，除了跳转这个动作以外，同时也传递了一个参数 cid，它传递的值是 002540。仿照这个例子，也传递一个值为 002540 的参数

cid，并且在 Servlet 里将参数传递的值输出到网页上。

图 3.9　百度首页跳转糯米网传递参数

目标：学习使用 Servlet，读取用户通过超级链接提交的信息。

工程名：eg0303。

用到的文件如表 3.3 所示。

表 3.3　eg0303 用到的文件及文件说明

文件名	说明
index. html	创建工程时自动创建。为项目首页。页面上有一个超级链接，用于完成跳转及传递参数
ShowCidServlet. java	手动创建的 Servlet，用来取得用户通过超级链接提交的 cid 信息，并以网页形式显示

编程思路：先取得参数 cid 的值并赋给一个临时变量，将该变量的值输出。

打开 NetBeans IDE，新建一个名为 eg0303 的 Java Web 项目，具体过程略；新建名为 ShowCidServlet. java 的 Servlet，具体过程略。编写各部分代码如下（仅列出核心代码，大部分自动生成的代码略）：

index. html 代码片段

```
<!DOCTYPE html>
<html>
    <head>
        <title>通过超级链接提交 cid 的值</title>
        <meta charset = "UTF - 8">
        <meta name = "viewport" content = "width = device - width, initial - scale = 1.0">
    </head>
    <body>
        <a href = "ShowCidServlet?cid = 002540">向 ShowCidServlet 提交 cid</a>
```

```
    </body>
</html>
```

在 eg0302 中使用了请求对象的 3 个方法，getHeader(String name)、getHeaders(String name)、getHeaderNames()。在这个练习中将使用请求对象的 getParameter(String name)方法，如表 3.4 所示。这个方法适用于一个参数名对应一个参数值的情况。

表 3.4 javax.servlet.http.HttpServletRequest 定义的方法

方法名	返回值类型	适用情况
getParameter(String name)	String	取得某个请求参数的参数值，假如该参数没有值则返回""；假如该参数不存在则返回 null

ShowCidServlet.java 代码片段

```
protected void processRequest(HttpServletRequest request, HttpServletResponse response)
    throws ServletException, IOException {
        response.setContentType("text/html;charset = UTF - 8");
        try (PrintWriter out = response.getWriter()) {
            //为了区分参数 cid 和返回值,返回值刻意声明了新的引用名
            //通过请求对象获得变量 cid 对应的值
            String mycid = request.getParameter("cid");
            out.print(mycid);
    }
}
```

运行结果如图 3.10 所示。

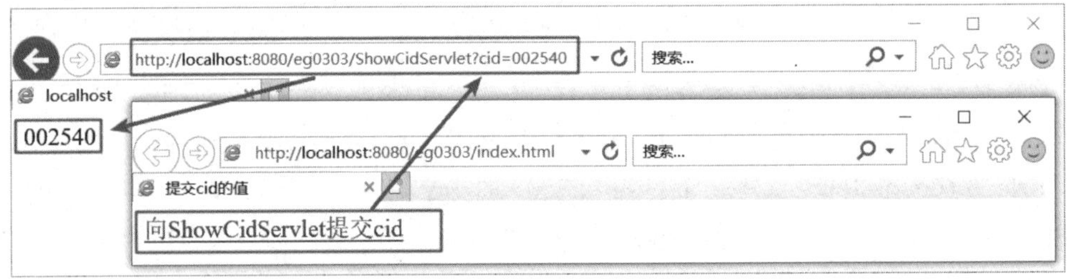

图 3.10 eg0303 运行结果

3.8 使用 Servlet 读取用户通过表单传送的信息

尽管表单内的控件多种多样，但是实际上抽象为两类即可。一类是类似于超级链接 cid = 002540 这种，一个变量名对应一个变量值，例如：文本框、密码框、单选按钮等。另一类是复选框这种，用户的选择不是固定的，可能一个也没选，也可能全都选了。例如下面代码中，参数 fruits 对应的值在集合{banana，apple，orange}中选取。

```
<input type="checkbox" name="fruits" value="banana"/> 香蕉
<input type="checkbox" name="fruits" value="apple"/> 苹果
<input type="checkbox" name="fruits" value="orange"/> 橙子
```

对于第一类，用的方法依然是请求对象的 getParameter(String name)方法。而对于第二类，则要用到请求对象的另一个方法 getParameterValues(String name)，如表 3.5 所示。

表 3.5 javax.servlet.http.HttpServletRequest 定义的方法

方 法 名	返回值类型	适 用 情 况
getParameterValues(String name)	String[]	如果某个参数可能有多个值，可用此方法一次取得全部的值并封装在一个 String 数组中；假如该参数不存在则返回 null

目标：学习使用 Servlet，读取用户通过表单提交的信息。

工程名：eg0304。

用到的文件如表 3.6 所示。

表 3.6 eg0304 用到的文件及文件说明

文 件 名	说 明
index.html	创建工程时自动创建。为项目首页。页面上有一个表单。表单内有一个文本框，一组复选框
ShowFormServlet.java	手动创建的 Servlet，用来取得用户通过表单提交的信息，并以网页形式显示

编程思路：依次取得文本框和复选框的值，并分别输出，中间以水平线分隔。

打开 NetBeans IDE，新建一个名为 eg0304 的 Java Web 项目，具体过程略；新建名为 ShowFormServlet.java 的 Servlet，具体过程略。编写各部分代码如下(仅列出核心代码，大部分自动生成的代码略)：

index.html 代码片段

```
<!DOCTYPE html>
<html>
    <head>
        <title>通过表单提交</title>
        <meta charset="UTF-8">
        <meta name="viewport" content="width=device-width, initial-scale=1.0">
    </head>
    <body>
        <form action="ShowFormServlet" method="post">
          <input type="text" name="msg" value="" />
          <br />
          <input type="checkbox" name="fruits" value="banana" />香蕉
          <input type="checkbox" name="fruits" value="apple" />苹果
          <input type="checkbox" name="fruits" value="orange" />橙子
          <br />
```

```
            <input type="submit" value="提交" />
        </form>
    </body>
</html>
```

ShowFormServlet.java 代码片段

```
protected void processRequest(HttpServletRequest request, HttpServletResponse response)
    throws ServletException, IOException {
        response.setContentType("text/html;charset=UTF-8");
        try (PrintWriter out = response.getWriter()) {
            //获得 msg 的值并输出
            String msg = request.getParameter("msg");
            out.print(msg);
            out.print("<hr />");
            //获得 fruits 的值,如果用户未做选择,提示"未做选择"
            //否则遍历输出
            String[] fruits = request.getParameterValues("fruits");
            if (fruits == null) {
                out.print("未做选择");
            } else {
                for (String fruit : fruits) {
                    out.print(fruit + "<br />");
                }
            }
        }
}
```

运行结果如图 3.11 所示。

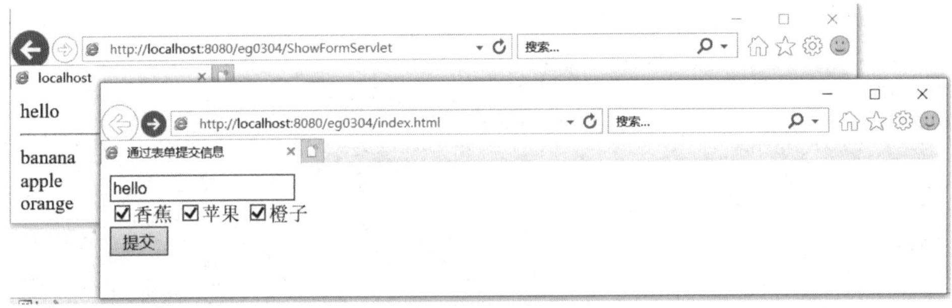

图 3.11 eg0304 运行结果

3.9 处理表单提交的中文乱码问题

重新运行 eg0304,在文本框中输入中文信息后提交,会发现尽管输出了信息,但是信息未能正确显示,出现了“乱码”,如图 3.12 所示。

图 3.12 eg0304 运行出现乱码

追本溯源，乱码的产生主要是因为数据在传递过程中编码格式不完全一致。但是如果细分，中文乱码的产生原因很多，根据不同的成因也有不同的解决办法，本节只介绍如何处理经过 post 方式发送请求，提交的信息中含有中文的情况。对于乱码问题的更多内容，感兴趣的读者可以通过搜索引擎继续学习。

处理经过 post 方式提交中文，从而产生乱码的问题，解决方案并不复杂。只需要在调用请求对象，通过请求对象获取请求参数值之前，添加一条语句就可以了。

```
request.setCharacterEncoding("UTF-8");
```

这条语句将请求对象的编码格式设置为 UTF-8，与响应对象设置的字符集相同。（并不是永远都设置成 UTF-8，需要视当前项目的具体情况而定，保持统一即可）。在 NetBeans IDE 中，默认的各字符编码均为 UTF-8，例如响应类型默认的字符编码也是 UTF-8。

```
response.setContentType("text/html;charset=UTF-8");
```

HTML 页面设置的编码也是 UTF-8。

```
<meta charset="UTF-8">
```

通常为了确保设置请求对象字符编码的语句绝对有效，需要把这条语句放在 Servlet 中的第一句，这样之后再通过请求对象获得参数值时，字符编码就肯定是 UTF-8 了。改写后的 ShowFormServlet 代码为：

ShowFormServlet.java 代码片段（处理中文乱码后）

```
protected void processRequest(HttpServletRequest request, HttpServletResponse response)
    throws ServletException, IOException {
        request.setCharacterEncoding("UTF-8");
        response.setContentType("text/html;charset=UTF-8");
        try (PrintWriter out = response.getWriter()) {
            //获得 msg 的值并输出
            String msg = request.getParameter("msg");
```

```
                out.print(msg);
                out.print("<hr />");
                //获得 fruits 的值,如果用户未做选择,提示"未做选择"
                //否则遍历输出
                String[] fruits = request.getParameterValues("fruits");
                if (fruits == null) {
                    out.print("未做选择");
                } else {
                    for (String fruit : fruits) {
                        out.print(fruit + "<br />");
                    }
                }
        }
    }
```

处理完乱码后再次输入数据测试程序,运行效果如图 3.13 所示。

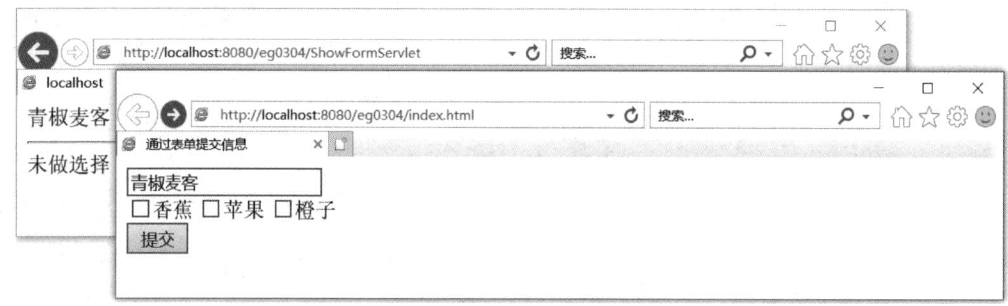

图 3.13 乱码处理完成后运行效果

3.10 对响应进行重定向

在实际应用中,取得用户的信息只是第一步,之后往往还要根据输入的信息,对用户的请求做出响应,定位到其他的资源(Servlet 或者网页)。如何解决这一问题呢?对响应进行重定向是一种解决方案。

对响应进行重定向,就是将响应重新定向到别的资源,相当于重新发送了一个 HTTP 请求。通过响应对象的 sendRedirect 方法可以实现响应的重定向,如表 3.7 所示。

表 3.7 javax.servlet.http.HttpServletResponse 定义的方法

方法名	返回值类型	适用情况
sendRedirect(String location)	void	将用户重定向至其他页面或网站。假如 location 不是以"/"开头,容器解析为相对于当前 URI。假如 location 以"/"开头,容器认为相对于当前 Servlet 所在容器的根目录。假如响应已经被提交了,调用这个方法将抛出 IllegalStateException 类型的异常

下面通过一个用户登录的示例演示通过响应对象的 sendRedirect 方法实现响应的重定向。

目标：学习使用响应对象，完成对响应进行重定向。

工程名：eg0305。

用到的文件如表 3.8 所示。

表 3.8 eg0305 用到的文件及文件说明

文件名	说明
login. html	手动创建的页面，包含一个表单，表单中有文本框用来输入用户名，密码框用来输入密码，提交按钮用来提交登录信息
error. html	手动创建的页面，页面显示“登录失败，返回登录页”，其中返回登录页为链接文字，链接至 login. html
HandleLoginServlet. java	手动创建的 Servlet，用来取得用户提交的登录信息，如果用户名为 admin，密码为 1234，则输出“欢迎光临！”，否则将响应重定向至 error. html

打开 NetBeans IDE，新建一个名为 eg0305 的 Java Web 项目，具体过程略；新建名为 login. html 和 error. html 的 HTML 文件；新建名为 HandleLoginServlet. java 的 Servlet，具体过程略。编写各部分代码如下(仅列出核心代码，大部分自动生成的代码略)：

login. html

```
<!DOCTYPE html>
<html>
    <head>
        <title>用户登录</title>
        <meta charset = "UTF-8">
        <meta name = "viewport" content = "width = device - width, initial - scale = 1.0">
    </head>
    <body>
        <form method = "post" action = "HandleLoginServlet">
            用户名：<input type = "text" name = "userName" value = "" /><br />
            密码：<input type = "password" name = "userPass" value = "" /><br />
            <input type = "submit" value = "登录" />
        </form>
    </body>
</html>
```

error. html

```
<!DOCTYPE html>
<html>
    <head>
        <title>登录失败</title>
        <meta charset = "UTF-8">
        <meta name = "viewport" content = "width = device - width, initial - scale = 1.0">
    </head>
```

```
        <body>
            <div>登录失败<a href="login.html">返回登录页</a></div>
        </body>
    </html>
```

HandleLoginServlet.java 代码片段

```
protected void processRequest(HttpServletRequest request, HttpServletResponse response)
        throws ServletException, IOException {
    response.setContentType("text/html;charset=UTF-8");
    try (PrintWriter out = response.getWriter()) {
        String userName = request.getParameter("userName");
        String userPass = request.getParameter("userPass");
        if ("admin".equalsIgnoreCase(userName) && "1234".equals(userPass)) {
            out.print("欢迎光临");
        } else {
            response.sendRedirect("error.html");
        }
    }
}
```

当用户名和密码输入错误时,效果如图 3.14 所示,注意地址栏,可以看到用户在 login.jsp 输入登录信息后,经过 Servlet 处理,被重新定向到了 error.jsp。

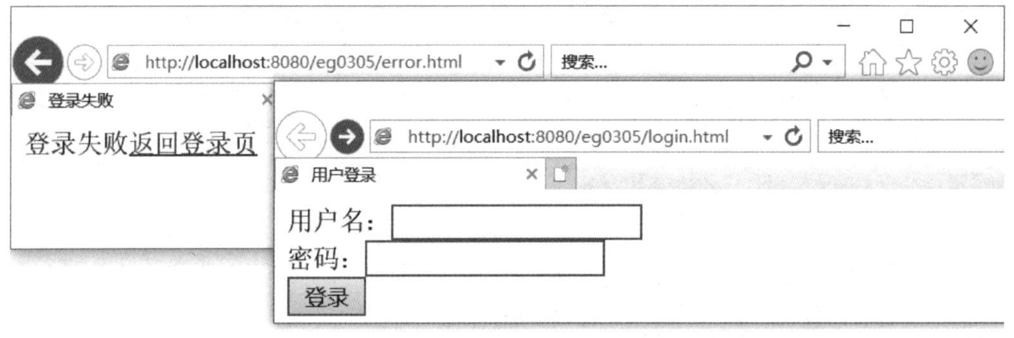

图 3.14 重新定向到了 error.html 效果

响应的重定向能够达到从一个资源跳转到另一个资源的目的,它类似于生活中拨打 114 的场景。例如说班级要搞团建,于是拨打 114 找某某饭店提前预订座位。这时 114 并不会帮你直接预订座位,而是把饭店的电话告诉你。你再拨打饭店电话才能由饭店帮你完成座位预定的需求。在这个过程中,你需要打两个电话,一个打给 114,一个打给饭店。

刚刚完成的用户登录的例子,用户最终提交了两次请求,一次是通过表单的 action 属性设置的,提交给 HandleLoginServlet;一次是由 Servlet 中的重定向语句提交的,提交给 error.html。

3.11 使用请求转发器(RequestDispatcher)转发请求

除了重定向响应以外，请求转发器的两个方法也可以实现在资源之间跳转，如表3.9所示。响应的重定向和请求转发器的方法适用场合不同。

请求转发器(javax.servlet.RequestDispatcher)是一个接口，可以通过请求对象得到一个RequestDispatcher对象，以下代码通过请求对象获得了一个转发器对象，该转发器可以将请求转发至index.html。

```
request.getRequestDispatcher("index.html");
```

通过转发器对象的相应方法可以实现请求的转发。

表3.9 javax.servlet.RequestDispatcher 定义的方法

方法名	返回值类型	适用情况
forward(ServletRequest request, ServletResponse response)	void	将请求从一个Servlet转发给服务器上其他的Servlet、JSP或者HTML
include(ServletRequest request, ServletResponse response)	void	在响应中包含其他资源的内容(如Servlet、JSP或者HTML)

注意：include()方法与forward()方法非常类似，唯一的不同在于：利用include()方法将HTTP请求转送给其他Servlet后，被调用的Servlet虽然可以处理这个HTTP请求，但是最后的主导权仍然是在原来的Servlet。换言之，被调用的Servlet如果产生任何HTTP响应，将会并入原来的HttpResponse对象。

下面通过用户登录样例的另一种实现来学习如何转发请求实现资源间的跳转。这个样例与eg0305有两个主要的区别：一是为了避免中文字符的乱码问题，网页文件不是HTML文档，而是JSP文档。二是实现的逻辑略有变化，输入的用户名密码正确时，显示index.jsp的内容；用户名和密码错误时，提示错误信息，并显示login.jsp页面内容。

目标：学习使用转发器对象(RequestDispatcher)的两个方法，实现对请求进行转发，并比较两个方法的异同。

工程名：eg0306。

用到的文件如表3.10所示。

表3.10 eg0306用到的文件及文件说明

文件名	说明
login.jsp	手动创建的JSP页面，包含一个表单，表单中有文本框用来输入用户名，密码框用来输入密码，提交按钮用来提交登录信息
index.jsp	手动创建的JSP页面(为了避免引起混淆，请将index.html文件删除)。页面显示“欢迎光临！”
HandleLoginServlet.java	手动创建的Servlet，用来取得用户提交的登录信息，如果用户名为admin，密码为1234，则转发至index.jsp，否则输出“用户名或密码错误，请重新输入！”并显示登录页(login.jsp)内容

打开 NetBeans IDE,新建一个名为 eg0306 的 Java Web 项目,具体过程略;删除 index.html 文件;新建名为 login.jsp、index.jsp 的 JSP 文件;新建名为 HandleLoginServlet.java 的 Servlet,具体过程略。编写各部分代码如下(仅列出核心代码,大部分自动生成的代码略)。新建 JSP 文件的过程如图 3.15 所示。

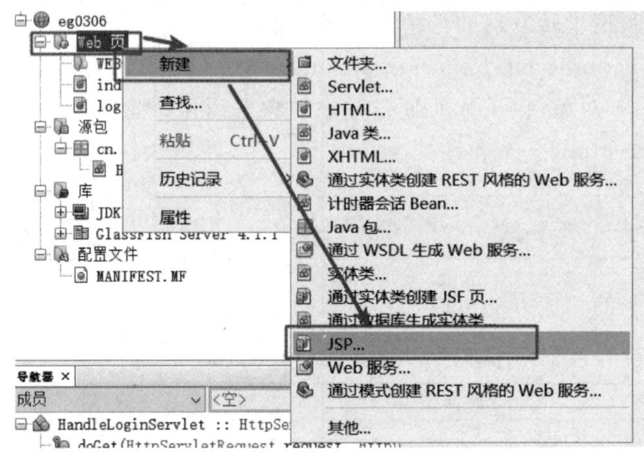

图 3.15 新建 JSP 文件

login.jsp

```
<%@page contentType="text/html" pageEncoding="UTF-8"%>
<!DOCTYPE html>
<html>
    <head>
        <meta http-equiv="Content-Type" content="text/html; charset=UTF-8">
        <title>用户登录</title>
    </head>
    <body>
            <h1>用户登录</h1>
            <form method="post" action="HandleLoginServlet">
                用户名:<input type="text" name="userName" value="" /><br />
                密码:<input type="password" name="userPass" value="" />
                <br />
                <input type="submit" value="登录" />
            </form>
    </body>
</html>
```

index.jsp

```
<%@page contentType="text/html" pageEncoding="UTF-8"%>
<!DOCTYPE html>
<html>
    <head>
        <meta http-equiv="Content-Type" content="text/html; charset=UTF-8">
```

```
        <title>首页</title>
    </head>
    <body>
            <h1>欢迎光临!</h1>
    </body>
</html>
```

HandleLoginServlet.java 代码片段

```
protected void processRequest(HttpServletRequest request, HttpServletResponse response)
    throws ServletException, IOException {
    response.setContentType("text/html;charset = UTF - 8");
        try (PrintWriter out = response.getWriter()) {
            String userName = request.getParameter("userName");
            String userPass = request.getParameter("userPass");
            if ("admin".equalsIgnoreCase(userName) && "1234".equals(userPass)) {
                out.print("用户名、密码正确");              //本行输出不可见
              RequestDispatcher rd = request.getRequestDispatcher("index.jsp");
                rd.forward(request, response);
            } else {
                out.print("用户名、密码错误,请重新输入");  //本行输出可见
              RequestDispatcher rd = request.getRequestDispatcher("login.jsp");
                rd.include(request, response);
            }
    }
}
```

请注意代码中单行注释部分，通过两条输出语句效果的不同，比较 forward 方法和 include 方法的不同。

```
out.print("用户名、密码正确");                  //本行输出不可见
out.print("用户名、密码错误,请重新输入");       //本行输出可见
```

当用户名、密码输入正确时，效果如图 3.16 所示，注意地址栏，可以看到用户在 login.jsp 输入正确的登录信息后，在 Servlet 中处理时，请求被转发到了 index.jsp(因为显示出了 index.jsp 的内容)，但是地址栏并没有变化，仍然是对 HandleLoginServlet 进行的请求。

当用户名、密码输入错误时，效果如图 3.17 所示，注意地址栏，可以看到用户在 login.jsp 输入登录信息后，在 Servlet 中处理时，请求被转发到了 login.jsp(因为显示出了 login.jsp 的内容)，但是地址栏并没有变化，仍然是对 HandleLoginServlet 进行的请求。

使用请求转发器对请求进行转发，类似于生活中拨打 120 的场景。假设某一天，有人意外受伤了，或者得了急病需要治疗，这时需要拨打 120 急救电话。假如 120 急救中心接到电话后只是告诉你对不起，我处理不了，请找某某医院。那么这种情况就相当于响应的重定向，问题并没有得到最终的解决，你被重定向到另一个部门了，会由另一个部门解决。

好在现实生活中不是这样，120 急救中心通常会了解病人的具体情况、地址等信息，然

图 3.16 用户名和密码输入正确时的运行效果

图 3.17 用户名和密码输入错误时的运行效果

后根据就近、就急、病人家属意愿和专业对应等原则，指挥调度 120 急救车开展医疗救治工作。这样，从病人的角度看，好像只是呼叫了 120 急救中心就得到了服务，但实际处理中既涉及了急救中心的接警部门，也涉及了医院及医护人员，经过了多个部门，这种情况就相当于通过请求转发器对请求进行转发。尽管只是一次请求，但是该请求被转发到了好几个部门，多个部门协同解决了这一请求。

刚刚完成的改版之后的用户登录的例子，当用户提交的用户名和密码正确时，请求 HandleLoginServlet。在 HandleLoginServlet 中，请求转发器根据转发条件，将请求通过 forward 方法转发给 index.jsp 页面。由 index.jsp 页面负责生成响应，完成了资源间的跳转，但是地址栏没变，是一次请求而不是两次。如图 3.16 所示。

当用户提交的用户名和密码错误时，请求 HandleLoginServlet。在 HandleLoginServlet 中，

请求转发器根据转发条件，将请求通过 include 方法转发给了 login.jsp 页面。依然由 HandleLoginServlet 负责生成响应，login.jsp 页面的内容被包含到最终的响应中，也完成了资源间的跳转，地址栏没变，也是一次请求而不是两次。如图 3.17 所示。

注意：在本节不必太过于纠结到底该使用 forward 方法还是 include 方法。在第 10 章中可以看到，当可用的技术手段丰富起来时，选择方案也更加多样化。在 10.3 节 MIS 第 1 版实现中，因为结合使用了第 8 章的 EL 技术，样例中并没有使用 include 方法，只使用 forward 方法和 EL 就完成了 eg0306 类似的功能。

3.12　Servlet 的生命周期

在本章前面的练习中，已经使用了 Servlet，并且通过 Servlet 完成了一些基本的任务。那么，你是否想过这个问题：Servlet 并没有主方法，也从来没有见过用 new someServlet() 之类的代码去实例化一个对象出来，那么这个 Servlet 对象是由谁来实例化、初始化、使用和销毁的呢？

Servlet 是运行在服务器端的一种组件，它由 Servlet 容器来实例化、初始化、使用和销毁。每当有用户对该 Servlet 发出请求时，容器就实例化一个对象，然后启动一个线程为该用户服务，当另外有用户请求同一个 Servlet 时，就再启动另一个线程提供服务，所以有时候会说 Servlet 是"单实例，多线程"的。因为实例化以及与线程相关的任务都由 Servlet 容器来完成，所以程序员就可以专注于其他的业务逻辑部分了。回想一下前面的操作，是不是只要在 processRequest 方法中编程就可以了？

那么容器又是如何实例化、初始化、使用和销毁 Servlet 对象的呢？下面先看一个代码片段，在 NetBeans IDE 中打开 eg0306 这个工程，再打开 HandleLoginServlet.java 这个文件，可以看到这样的代码：

```
@WebServlet(name = "HandleLoginServlet", urlPatterns = {"/HandleLoginServlet"})
    public class HandleLoginServlet extends HttpServlet {
        //代码略
    }
```

（中间略去的代码是 processRequest、doGet、doPost 等几个方法，因为这一次重点不是这几个方法）。可以看到 HandleLoginServlet 是继承自 HttpServlet。按住 Ctrl 键，同时在 HttpServlet 上单击，可以看到 HttpServlet 的代码。

```
public abstract class HttpServlet extends GenericServlet {
    //代码略
}
```

可见 HttpServlet 又继承自 GenericServlet。按住 Ctrl 键，同时在 GenericServlet 上单击，还可以看到 GenericServlet 的代码。

```
public abstract class GenericServlet implements Servlet, ServletConfig, Serializable{
    //代码略
}
```

可见 GenericServlet 实现了 3 个接口。其中一个接口是 Servlet。按住 Ctrl 键,同时在 Servlet 上单击,可以看到 Servlet 的代码。

```
public interface Servlet {
    public void init(ServletConfig config) throws ServletException;
    public ServletConfig getServletConfig();
    public void service(ServletRequest req, ServletResponse res) throws ServletException,
IOException;
    public String getServletInfo();
    public void destroy();
}
```

在这段代码中就有答案了。Servlet 的生命周期分为四个阶段: 实例化、初始化、对外服务、销毁。

(1) 实例化时调用的是 Servlet 的构造方法。

(2) 初始化时调用的是 init()方法,这个方法仅在 Servlet 首次载入时执行一次,并不是每次请求都要调用,这就是前面说的"单实例,多线程"。在 GenericServlet 中提供了两种初始化的方式: 常规初始化和由初始化参数控制的初始化。

(3) 对外提供服务时调用的是 service()方法,这个方法在新线程中由服务器为每个请求而调用,但是并不直接对这个方法编程而是针对不同的请求方式调用相应的 doXxx()方法,例如前面提到过的 doGet()方法和 doPost()方法等。

(4) 销毁时调用 destroy()方法,该方法在服务器删除 Servlet 的实例时调用。并不是每次请求后都调用这个方法。

其实即使不知道这些,也不影响使用 Servlet 来编程,但是多了解一下也好,可以让我们对 Servlet 理解得更为透彻。下面通过一个例子来证明。

目标: 理解 Servlet 的生命周期。

工程名: eg0307。

打开 NetBeans IDE,新建一个名为 eg0307 的 Java Web 项目,具体过程略; 新建名为 NewServlet.java 的 Servlet,只设置包名为 cn.edu.djtu,其他设置取默认值。

在现有的 NewServlet.java 代码中,有四个方法 doGet、doPost、getServletInfo、processRequest,其中 doGet、doPost 均调用了 processRequest 方法,processRequest 方法是由 NetBeans 根据模板自动生成的。在 22 行代码的空白处,如图 3.18 所示,按快捷键"Ctrl+\"就可以根据提示选择要覆盖或者生成的方法。先生成一个构造方法 NewServlet(),然后再次在空白的地方按快捷键"Ctrl+\"——按 I 键,就可以看到提示覆盖 init 方法,init 方法有两个,一个是常规的初始化方法,一个是通过 ServletConfig 对象初始化的方法,选择无参的就好,如图 3.19 所示。

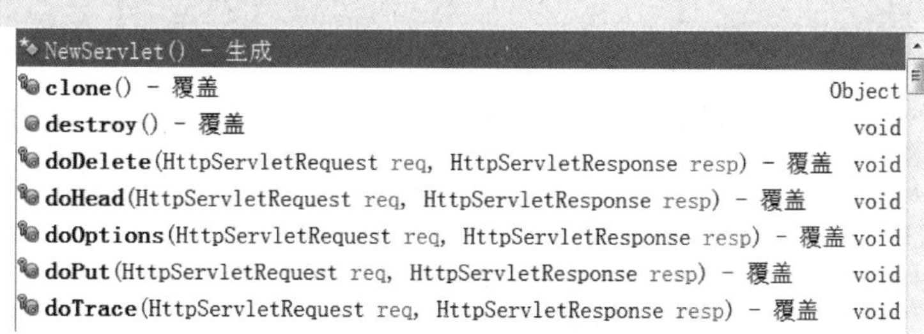

图 3.18 生成构造方法

图 3.19 覆盖 init 方法

按照类似的操作，可以覆盖 service()方法，service()方法也有两个，覆盖参数类型为 HttpServletRequest 和 HttpServletResponse 的就可以了，如图 3.20 所示。

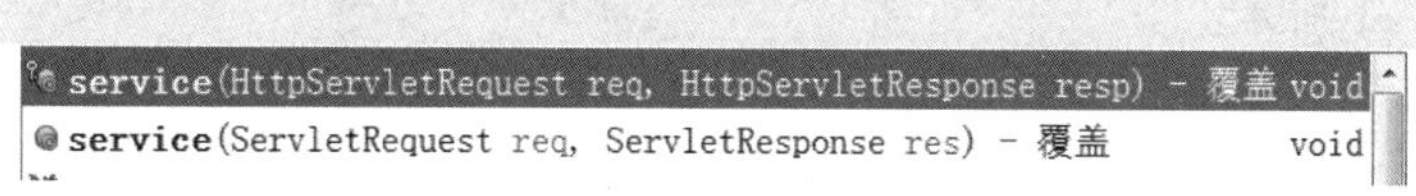

图 3.20 覆盖 service()方法

最后用类似的操作覆盖 destroy()方法，图略。然后在每一个方法中添加一条控制台的输出语句，当该方法被调用时，会在控制台输出，这样就可以知道各个方法的执行顺序了。代码中省略了 doGet、doPost、getServletInfo 三个方法的代码，因为 doGet、doPost 方法调用的是 processRequest 方法，全部的代码如下：

NewServlet.java

```
@WebServlet(name = "NewServlet", urlPatterns = {"/NewServlet"})
public class NewServlet extends HttpServlet {
    public NewServlet() {
        System.out.println("调用构造方法");
    }
    @Override
    public void init() throws ServletException {
        System.out.println("调用 init 方法");
        super.init();
    }
    @Override
    protected void service(HttpServletRequest req, HttpServletResponse resp) throws
ServletException, IOException {
        System.out.println("调用 service 方法");
        super.service(req, resp);
```

```
    }
    @Override
    public void destroy() {
        System.out.println("调用 destroy 方法");
        super.destroy();
    }
    protected void processRequest(HttpServletRequest request, HttpServletResponse response)
throws ServletException, IOException {
    response.setContentType("text/html;charset = UTF-8");
        try (PrintWriter out = response.getWriter()) {
            System.out.println("调用 processRequest 方法");
        }
    }
    //省略了 doGet、doPost、getServletInfo 三个方法的代码
    //因为 doGet、doPost 均调用 processRequest 方法
}
```

运行这个Servlet,观察输出窗口的输出情况可以知道：先调用了构造方法,然后调用了初始化方法 init(),接着调用了提供服务的方法 service(),service()方法又调用了 processRequest 方法,但是没有看到“调用 destroy 方法”字样,说明 Servlet 对象目前还没有被销毁,如图 3.21 所示。

图 3.21 第 1 次访问时各方法执行情况

接着模拟另一个用户的访问,看看运行情况。另开一个浏览器窗口,将刚才访问的 URL“http://localhost:8080/eg0307/NewServlet”复制粘贴到地址栏并访问。运行结果见图 3.22。

可见又调用了一次 service()方法和 processRequest()方法,没有调用构造方法和初始化方法 init()。由此可见只创建了一个 Servlet 对象。这就证明了前面说过的：Servlet 是“单实例,多线程”的。也证明了 Servlet 的生命周期的阶段划分。

在上面操作的基础上,再做一点修订,将 service()方法中的 super.service(req, resp);语句注释掉然后保存,代码如下：

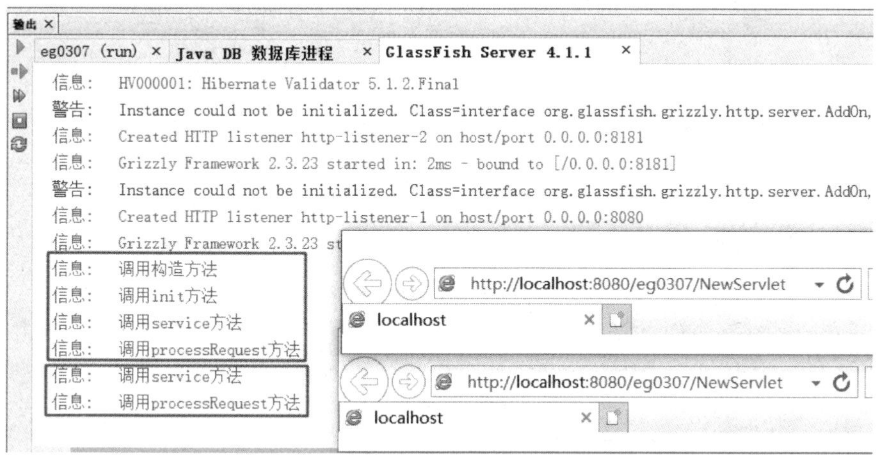

图 3.22 第 2 次访问时各方法执行情况

```
@Override
protected void service ( HttpServletRequest req, HttpServletResponse resp ) throws
ServletException, IOException {
    System.out.println("调用 service 方法");
    //super.service(req, resp);
}
```

由于代码改变了，所以 GlassFish Server 重新进行热部署，此时当前 Servlet 对象需要销毁，所以控制台能看到“调用 destroy 方法”字样，说明调用了 destroy 方法，如图 3.23 所示。

图 3.23 destroy 方法调用情况

此时，再一次运行 NewServlet，这时候服务器会重新启动，重新部署，也就意味着会重新实例化对象，初始化对象，对外提供服务，等待销毁，也许大家会觉得和上次一样，但是请看运行情况，如图 3.24 所示。

覆盖 service()方法后的运行结果。可以看到，调用了构造方法，调用了初始化方法 init()，调用了 service()方法。但是，没有调用 processRequest()方法。为什么呢，因为将下面这条

```
输出 ×
eg0307 (run) × Java DB 数据库进程 × GlassFish Server 4.1.1 ×
信息:  visiting unvisited references
信息:  visiting unvisited references
信息:  visiting unvisited references
信息:  Loading application [eg0307] at [/eg0307]
信息:  eg0307 was successfully deployed in 414 milliseconds.
信息:  visiting unvisited references
信息:  visiting unvisited references
信息:  visiting unvisited references
信息:  Loading application [eg0307] at [/eg0307]
信息:  eg0307 was successfully deployed in 340 milliseconds.
信息:  调用构造方法
信息:  调用init方法
信息:  调用service方法
```

未调用processRequest方法

图 3.24　覆盖 service 方法后的运行结果

语句注释掉了。

```
//super.service(req, resp);
```

在父类的方法中，service 会根据请求的类型，自动调用 doXxx 方法，而注释掉之后，service 就不会自动去调用相应的 doXxx 方法了，当然 processRequest()方法也就没有了执行的机会了。

通过以上的讲解，相信大家对 Servlet 的生命周期应该有了一个清晰的认识。那么接下来看一下 Servlet 的部署，看看容器是如何找到 Servlet 的。

3.13　Servlet 的部署

在早期的开发中，部署 Servlet 时是一定要用到 web.xml 文件的。web.xml 文件中含有 Servlet 的部署描述符，一共是两组标记：<servlet></servlet>和<servlet-mapping></servlet-mapping>，如图 3.25 所示。

以图 3.25 中程序为例，对于外部的用户而言，只需要知道访问的时候按<url-pattern></url-pattern>中设置的"/NewServlet"这种形式就可以访问资源了，至于这个资源是什么，在哪里，其实是一无所知的。容器在接到这种请求的时候就在配置文件的<servlet-mapping></servlet-mapping>中查找，看看这个<url-pattern>/NewServlet</url-pattern>对应的<servlet-name></servlet-name>是什么，于是找到了<servlet-name>NewServlet</servlet-name>；接下来就在<servlet></servlet>中查找形如"<servlet-name>NewServlet</servlet-name>"的标记，于是找到了<servlet-class>cn.edu.djtu.servlet.NewServlet</servlet-class>，然后将其实例化。

为了进一步理解 Servlet 的部署，可以分别做如下几个尝试以作对比：

(1) 将<url-pattern>/NewServlet</url-pattern>改为

<url-pattern>/test.html</url-pattern>，然后运行 index.jsp 页面，在运行出来的浏览

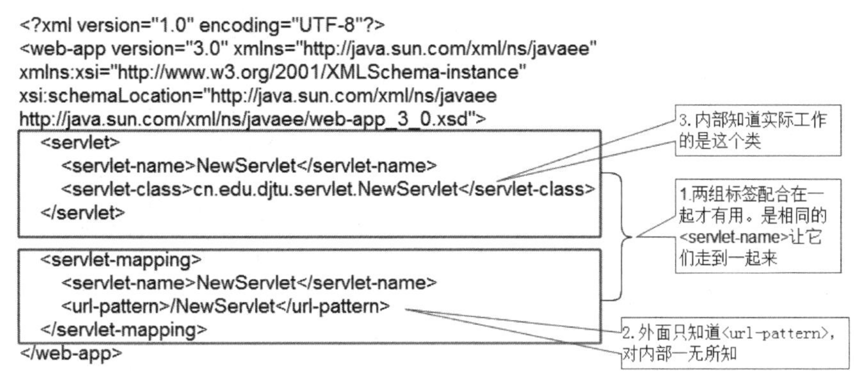

图 3.25　web.xml 中的 Servlet 配置

器地址栏中将 index.jsp 改成 test.html，会发现实际访问的是 NewServlet。

(2) 将两个< servlet-name > NewServlet </servlet-name >都改为< servlet-name > test </servlet-name >，然后运行 NewServlet，仍然能运行，说明这个值也是可以单独配置的。

从维护的角度看，通常会取默认值，不做修订，有特殊需要时，再个别设置某个 Servlet 就行了。

在新的 Servlet 规范中，已经建议使用 Annotation(注解)，而不是配置文件来完成 Servlet 部署工作。形式如下：

```
@WebServlet(name = "NewServlet", urlPatterns = {"/NewServlet"})
public class NewServlet extends HttpServlet {
    //代码略
}
```

不难理解其中 name = "NewServlet"作用等价于< servlet-name > test </servlet-name >；urlPatterns = {"/NewServlet" }作用等价于< url-pattern > /NewServlet </url-pattern >。

其实无论哪一种部署形式，都能顺利访问到 Servlet。对于 Servlet 的部署，由于篇幅所限，还有很多细节没有阐明，例如：一个 Servlet 可不可以有多个 urlPattern 呢？可不可以在应用一启动时就实例化一个 Servlet 呢？等等。如果想要深入了解，请查阅其他材料，对于大多数开发而言，部署形式取默认值就可以了。

3.14 本章回顾

本章简要介绍了 HTTP/HTTPS 的知识及其运行模型；介绍了 Servlet 的定义和作用；通过示例讲解了 Servlet 中需要覆写的方法；讲解了如何生成服务器响应；如何读取请求报头信息；如何读取用户提交的信息；如何实现资源间的跳转。最后讲解了 Servlet 的生命周期，如图 3.26 所示。

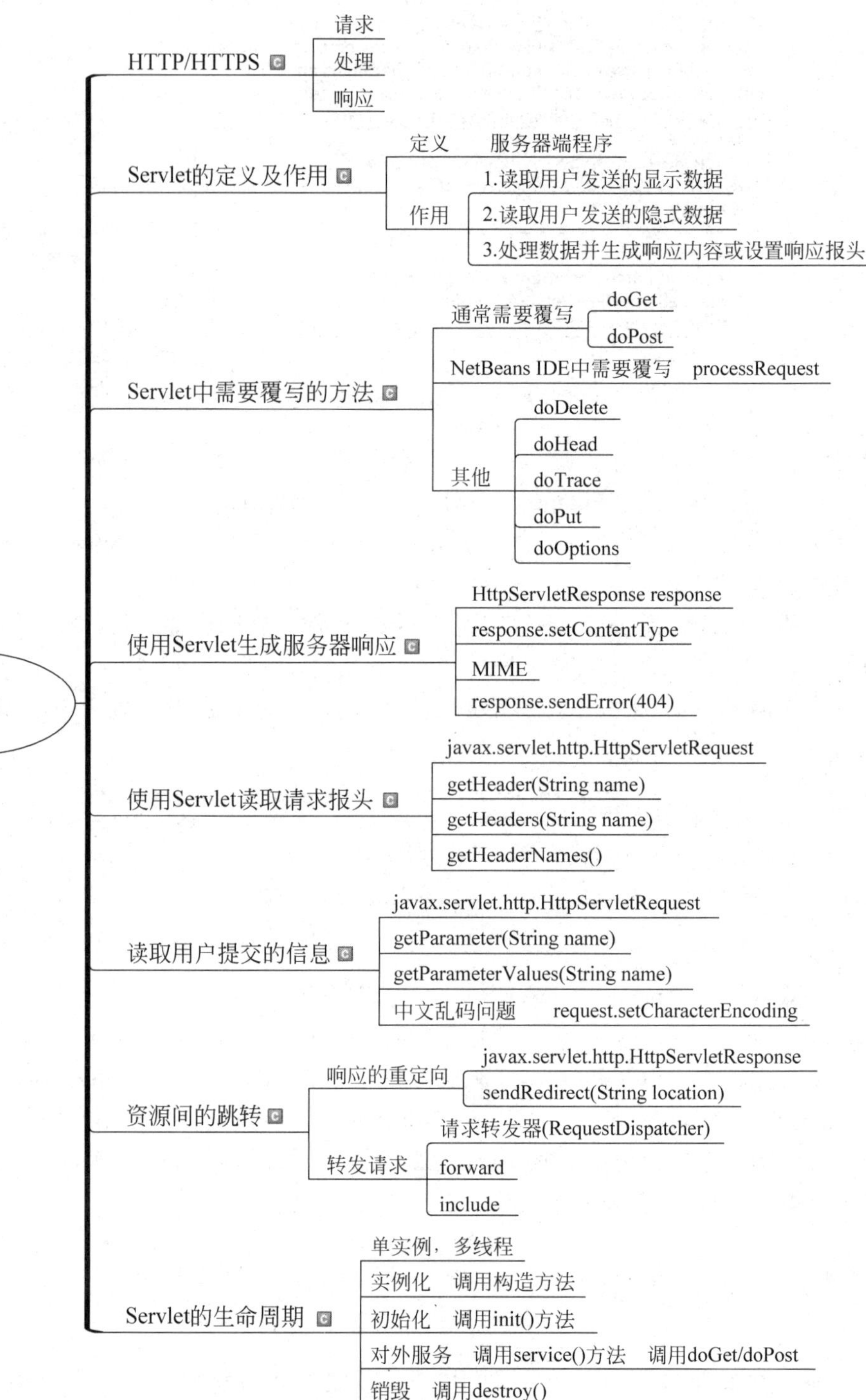

图 3.26 第 3 章内容结构图

3.15　课后习题

1. HTTP/HTTPS 的工作模型。
2. Servlet 的主要作用是什么？
3. Servlet 中一般需要覆写哪些方法？
4. 在 NetBeans IDE 中编写 Servlet 需要覆写哪个方法？
5. 如何设置响应类型？
6. 什么是 MIME？
7. 请求对象与获取请求报头相关的方法有哪些？
8. 读取用户提交的信息，一般要用到请求对象的哪些方法？对应何种场景？
9. 如何实现资源间的跳转？有几种方式？分别是什么？
10. 使用响应对象实现资源间的跳转，需要用到哪些/哪个方法？
11. 使用请求转发器对象实现资源间的跳转，需要用到哪些/哪个方法？
12. 比较响应的重定向与请求转发器的转发之间的不同。
13. 请求转发器的 forward 方法与 include 有什么区别与联系？

第4章 Servlet会话跟踪

学习目标：

通过本章的学习，你应该：

- 理解什么是会话
- 理解为什么需要会话跟踪
- 了解常用的会话跟踪技术
- 掌握会话对象的常用方法
- 熟练使用会话对象实现会话跟踪
- 理解浏览器端会话和服务器端会话的区别
- 掌握如何废弃会话
- 掌握 URL 重写的方法

4.1 会话概述

1. 什么是会话

生活中两个人聊天，大多是一问一答，有来有往。这个对话的过程可以称为一次会话。如图 4.1 所示。

在 Web 应用程序中，情况与此十分类似，如图 4.2 所示。

打开浏览器访问某个网站，标志着会话开始。在这个网站继续浏览其他页面，相当于不断地发送 HTTP 请求并接受 HTTP 响应。最后关掉浏览器或者单击“退出”则会话结束。

当然也有例外的情况，如果客户端发送一次请求后就不再有任何请求动作，则通常经过一段时间后，服务器端从提高运行效率出发，会终止这次会话。这就好像你对面坐着一个人，你和他说了一句话或几句话之后，他开始发呆，对你不理不睬，你当然不会无限期等下去，多数人会等待一段时间，然后就不再理他，终止这次会话。在 Web 应用程序中也是这样，服务器端可以设置等待超时的时间，过了这个期限，而客户端仍没有任何动作，服务器就会终止本次会话。

综上，可以这样简单地理解会话：从用户打开浏览器并向服务器发送请求开始，一直到用户关闭浏览器或者单击客户端程序中的“退出”按钮为止，这期间的连续调用过程称为一次会话。会话期间，可能会涉及多次“请求——响应”的调用。

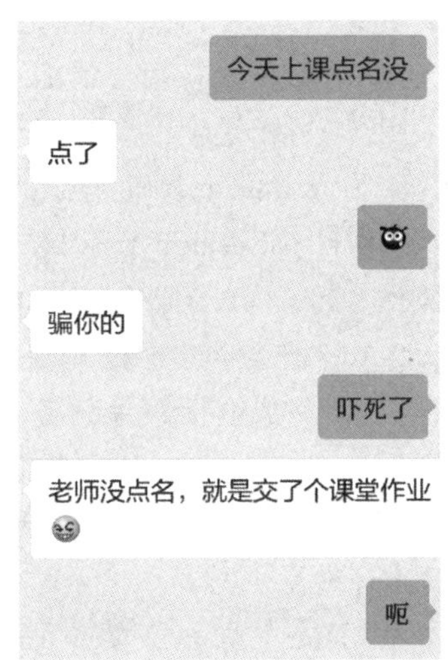

图 4.1 生活中的会话示例

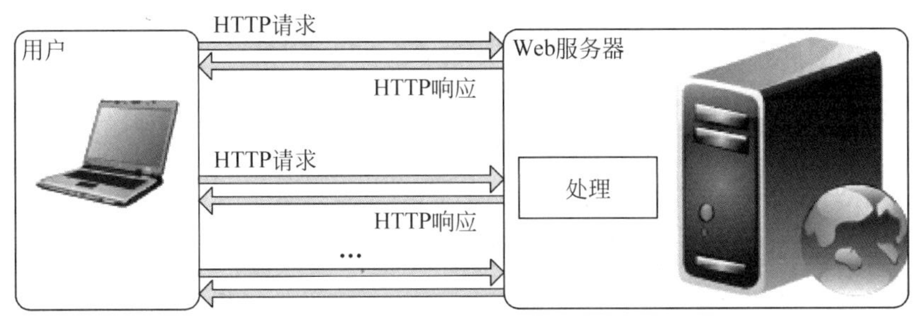

图 4.2 Web 应用程序中的会话

2. 为什么需要会话跟踪

为什么需要会话跟踪还得从 HTTP 说起。HTTP 是一个“无状态”的协议，所谓的“状态”指的是一个协议能够记忆用户和他的请求的能力。在“无状态”的 HTTP 中，服务器对每次的请求都一视同仁，总是根据当次提交的信息来给出响应，每一次请求对服务器来说都是新的请求。

HTTP 的这种“无状态”带来了一个问题，在大多数 Web 应用程序中，服务器必须有能力跟踪用户的会话，而一次会话中间可能会涉及多次请求，但是使用 HTTP 使得服务器很“健忘”，每次都认为是新的请求而不会把多个请求与某个用户关联起来。

因此需要采用某种会话跟踪技术，记住用户和他的请求，从而将单个无状态的 HTTP 请求转换为整体有状态的 Web 应用。

3. 常用的会话跟踪技术

会话跟踪技术的原理并不复杂。当服务器收到一个客户端第一次发来的请求时，服务器可以生成一个唯一的标识符，即会话 ID。客户端在随后发送的每一次请求中都必须包含这个会话 ID。在服务器端通过该 ID 来识别用户的请求属于哪一个会话，这样就把用户和该用户的一系列请求关联起来了。

在实际应用中既可以选择自己动手生成并管理会话ID,例如,采用Cookie、URL重写、隐藏的表单字段等技术,也可以采用Servlet提供的会话跟踪API。

自己动手实现会话跟踪技术,面临大量的重复性工作,而Servlet提供了一种简单的会话跟踪解决方案:HttpSession。利用HttpSession会话接口可以让程序员在更高的层次上解决问题,从而摆脱底层的烦琐操作。Servlet容器负责接口的实现,在需要进行会话跟踪的时候,通过调用相应的方法,Servlet容器会创建会话对象,通过会话对象的相应方法可以存取用户的会话ID或者其他信息,从而完成会话跟踪任务。

HttpSession的底层实现是基于Cookie或URL重写技术的,当客户端允许使用Cookie时,则优先使用Cookie进行会话追踪,只是不再需要直接操纵Cookie对象,如果客户端禁用Cookie,则需要选择对URL进行重写。

在实际的Web应用程序中,程序员一般只需要掌握HttpSession接口常用的方法即可。

4.2 获得与当前用户相关联的会话对象

通过调用请求对象的getSession()方法可以访问到HttpSession对象。getSession()方法是一个被重载了的方法:getSession()和getSession(boolean create)如表4.1所示。

表4.1 javax.servlet.http.HttpServletRequest定义的方法

方 法 名	返回值类型	适 用 情 况
getSession()	HttpSession	假如会话对象存在就返回与当前用户相关的会话对象;假如会话对象不存在就创建一个会话对象
getSession(boolean create)	HttpSession	当参数取值true时,逻辑与getSession()方法相同。当参数取值false时,假如会话存在就返回与当前用户相关的会话对象;假如会话不存在则返回null

那么一般什么时候使用getSession(true),什么时候使用getSession(false)呢?如果不关心会话对象中已经存入的信息,那么使用getSession(true)比较恰当。如果关心会话对象中的已有的信息,例如:用户访问电子商务网站时,在访问过程中,如果仅仅是想输出购物车中的商品信息,那么使用getSession(false)比较恰当。因为在业务逻辑上,如果购物车中已经存在商品,那么应该访问已有的会话对象并读取相关信息输出;如果没有商品,应该提示用户尚未选购任何商品。此时采用getSession(false),通过返回值是否为null即可完成业务逻辑,没必要非得创建一个对象,毕竟创建对象会产生一定的开销。

通过HttpSession对象的getId()方法可以取得服务器分配给该会话的ID。通过比较ID,可以判断会话对象是否是同一个,如表4.2所示。

表4.2 javax.servlet.http.HttpSession定义的方法

方 法 名	返回值类型	适 用 情 况
getId()	String	返回服务器分配给该会话的ID

下面通过一个简单的示例演示如何通过 Servlet 来访问与当前用户相关联的会话对象。

目标：学习使用 HttpSession 对象，访问与当前请求相关联的会话对象。

工程名：eg0401。

用到的文件如表 4.3 所示。

表 4.3 eg0401 用到的文件及文件说明

文 件 名	说 明
index. html	创建工程时自动创建的文件，对其进行简单改写即可。用户用来提交信息的页面。包含一个表单，表单内有一个文本字段和一个提交按钮，信息提交给 SessionServlet1
SessionServlet1. java	手动创建的 Servlet，用来取得 index. html 页面用户提交的信息，并以网页形式显示。在这个 Servlet 中访问与当前用户相关联的会话，并显示会话 ID；此外，输出一个超级链接，链接到 SessionServlet2
SessionServlet2. java	手动创建的 Servlet，用来取得与当前请求相关联的会话，并显示会话 ID；尝试通过 request. getParameter(String name)方法取得 index. html 提交的信息

编程思路：

(1) index. html 页面提交信息给 SessionServlet1，这是第一次请求，请求方式为 post。

(2) SessionServlet1 提交信息给 SessionServlet2，这是第二次请求，请求方式为 get，因为是通过超级链接提交的。

(3) 比较两次输出的 ID 是否相同，如相同，则说明是同一用户。

(4) 比较两次输出的用户提交信息是否相同。思考为什么？

打开 NetBeans IDE，新建一个名为 eg0401 的 Java Web 项目，具体过程略；分别新建名为 SessionServlet1. java、SessionServlet2. java 的 Servlet，具体过程略。编写各部分代码如下(仅列出核心代码，大部分自动生成的代码略)：

index. html

```
<!DOCTYPE html>
<html>
    <head>
        <title>提交信息</title>
        <meta charset="UTF-8">
        <meta name="viewport" content="width=device-width, initial-scale=1.0">
    </head>
    <body>
        <form action="SessionServlet1" method="post">
            想说的话: <input type="text" name="msg" value="" />
            <input type="submit" value="提交" />
        </form>
    </body>
</html>
```

SessionServlet1.java 代码片段

```
protected void processRequest(HttpServletRequest request, HttpServletResponse response)
    throws ServletException, IOException {
        response.setContentType("text/html;charset = UTF - 8");
        try (PrintWriter out = response.getWriter()) {
            //取得并输出用户提交的信息
            String msg = request.getParameter("msg");
            out.print(msg);
            //输出水平线作为分隔以便更清楚地看到效果
            out.print("<hr />");
            //访问与当前请求相关联的会话对象,如果不存在,则创建一个
            HttpSession session = request.getSession();
            //获得会话对象的 ID
            String sessionId = session.getId();
            out.print(sessionId);
            //输出水平线作为分隔
            out.print("<hr />");
            //输出超级链接,注意 href 属性值的双引号""需要转义
            out.print("<a href = \"SessionServlet2\">链接到 SessionServlet2 </a>");
        }
    }
```

SessionServlet2.java 代码片段

```
protected void processRequest(HttpServletRequest request, HttpServletResponse response)
    throws ServletException, IOException {
        response.setContentType("text/html;charset = UTF - 8");
        try (PrintWriter out = response.getWriter()) {
            //尝试获得并输出用户提交的信息
            String msg = request.getParameter("msg");
            out.print(msg);
            //输出水平线
            out.print("<hr />");
            //访问与当前请求相关联的会话对象,如果不存在,则创建一个
            HttpSession session = request.getSession();
            String sessionId = session.getId();
            out.print(sessionId);
        }
}
```

程序运行结果如图 4.3 所示。

运行程序,从 index.html 开始。在 index.html 页面输入想说的话:"hello",index.html 提交的信息在 SessionServlet1 中被取到并输出,同时输出此次会话的会话 ID。通过输出的超级链接可以跳转到 SessionServlet2。

单击超级链接后跳转到 SessionServlet2,尝试输出用户在 index.html 页面提交的信息,未果,实际输出为 null;尝试输出会话 ID,成功。经比对与 SessionServlet1 输出的会话

图 4.3 获得与当前用户相关联的会话对象

ID 相同，说明能够识别是同一用户发起的请求，对会话的跟踪成功。

结论：

(1) 两次输出的 ID 相同，说明是同一用户，HttpSession 对象顺利实现会话跟踪任务。

(2) 不论会话过程中经历的是 get 请求还是 post 请求，HttpSession 对象均能实现会话跟踪任务。

(3) 两次取得并输出用户提交的信息，第一次顺利取出，第二次为 null，验证了 HTTP 是一个"无状态"的协议，无法跨越多个请求。

4.3 在会话对象中存入、读取和移除信息

在 eg0401 中，虽然通过 HttpSession 对象识别出了同一个用户，成功完成了会话跟踪，但是用户在 index.html 页面提交的信息在 SessionServlet2 却无法取到，因为这已经是第二次请求了。那么 HttpSession 对象有没有能力将这个信息保留下来直到会话消失呢？当然可以。

HttpSession 对象有自己内建的数据结构(散列表)，它可以存储任意数量的键(key)和与键相关联的值(value)。键(key)通常称为属性名，数据类型为 String 型；值(value)称为属性值，数据类型为 Object，所以取得其值后需要进行数据类型的强制转换。

HttpSession 对象提供了与属性操作相关的一系列方法，最主要的有 3 个：setAttribute

(String name,Object value)、getAttribute(String name)、removeAttribute(String name),涵盖了存入信息、读取信息和移除信息,如表 4.4 所示。

表 4.4 javax.servlet.http.HttpSession 定义的方法

方 法 名	返回值类型	适 用 情 况
setAttribute (String name, Object value)	void	将某个对象绑定至当前会话,成为其属性之一。如果该属性已经存在,就用新对象替换已存在的对象
getAttribute(String name)	Object	从当前对象中取出指定属性绑定的对象,如不存在该属性名则返回 null
removeAttribute(String name)	HttpSession	移除指定属性所绑定的对象,如果该属性不存在则什么也不做

下面通过一个简单的示例演示如何在会话对象中存入、读取和移除信息。

目标:学习使用会话(HttpSession)对象,在会话对象中存入、读取和移除信息。

工程名:eg0402。

用到的文件如表 4.5 所示。

表 4.5 eg0402 用到的文件及文件说明

文 件 名	说 明
index.html	创建工程时自动创建的文件,对其进行简单改写即可。用户用来提交信息的页面,包含一个表单,表单内有一个文本字段和一个提交按钮,信息提交给 SessionServlet1
SessionServlet1.java	手动创建的 Servlet,用来取得 index.html 页面用户提交的信息,并以网页形式显示。在这个 Servlet 中访问与当前请求相关联的会话,将用户提交的信息存入会话对象,并显示会话 ID;此外,输出一个超级链接,链接到 SessionServlet2
SessionServlet2.java	手动创建的 Servlet,用来取得与当前请求相关联的会话,从会话对象中读取之前存入的信息,输出该信息并显示会话 ID。此外,输出一个超级链接,链接到 SessionServlet3
SessionServlet3.java	手动创建的 Servlet,用来取得与当前请求相关联的会话,移除之前存入的信息。再次读取存入的信息并输出该信息,最后显示会话 ID

编程思路:

(1) index.html 页面提交信息给 SessionServlet1,在 SessionServlet1 中练习如何在会话对象中存入信息。

(2) 在 SessionServlet1 中通过超级链接跳转到 SessionServlet2,在 SessionServlet2 中练习如何读取会话对象中存入的信息。

(3) 在 SessionServlet2 中通过超级链接跳转到 SessionServlet3,在 SessionServlet3 中练习如何移除会话对象中存入的信息。

(4) 比较三次输出的 ID 是否相同,如相同,则说明是同一用户,操纵的是相同的会话对象。

打开 NetBeans IDE,新建一个名为 eg0402 的 Java Web 项目,具体过程略;分别新建名为 SessionServlet1.java、SessionServlet2.java、SessionServlet3.java 的 Servlet,具体过程

略。eg0402练习中逻辑结构及大部分代码与eg0401相同，所以可以右击“eg0401”项目，选择“复制”命令，修改项目名称为eg0402；编写各部分代码如下（仅列出核心代码，大部分自动生成的代码略）：

index. html

```
<!DOCTYPE html>
<html>
    <head>
        <title>提交信息</title>
        <meta charset = "UTF-8">
        <meta name = "viewport" content = "width = device - width, initial - scale = 1.0">
    </head>
    <body>
        <form action = "SessionServlet1" method = "post">
            想说的话: <input type = "text" name = "msg" value = "" />
            <input type = "submit" value = "提交" />
        </form>
    </body>
</html>
```

SessionServlet1. java 代码片段

```
protected void processRequest(HttpServletRequest request, HttpServletResponse response)
throws ServletException, IOException {
    response.setContentType("text/html;charset = UTF - 8");
    try (PrintWriter out = response.getWriter()) {
        //取得并输出用户提交的信息
        String msg = request.getParameter("msg");
        out.print(msg);
        //输出水平线作为分隔以便更清楚地看到效果
        out.print("<hr />");
        //访问与当前请求相关联的会话对象,如果不存在,则创建一个
        HttpSession session = request.getSession();
        //获得会话对象的 ID
        String sessionId = session.getId();
        //输出会话 ID
        out.print(sessionId);
        //将用户提交的信息 msg 的值(value)以 myMsg 命名(key)存入会话对象
        session.setAttribute("myMsg", msg);
        //输出水平线作为分隔
        out.print("<hr />");
        //输出超级链接,注意 href 属性值的双引号""需要转义
        out.print("<a href = \"SessionServlet2\">链接到 SessionServlet2 </a>");
    }
}
```

SessionServlet2. java 代码片段

```
protected void processRequest(HttpServletRequest request, HttpServletResponse response)
throws ServletException, IOException {
    response.setContentType("text/html;charset = UTF - 8");
    try (PrintWriter out = response.getWriter()) {
        //访问与当前请求相关联的会话对象,如果不存在,则创建一个
        HttpSession session = request.getSession();
        //输出会话 ID
        String sessionId = session.getId();
        out.print(sessionId);
        //输出水平线
        out.print("<hr />");
        //从会话对象中取出名(key)为 myMsg 的对象,并强制转换为 String 类型
        String msg = (String) session.getAttribute("myMsg");
        out.print(msg);
        //输出水平线
        out.print("<hr />");
        //输出超级链接,链接到 SessionServlet3,注意 href 属性值的双引号""需要转义
        out.print("<a href = \"SessionServlet3\">移除信息</a>");
    }
}
```

SessionServlet3. java 代码片段

```
protected void processRequest(HttpServletRequest request, HttpServletResponse response)
throws ServletException, IOException {
    response.setContentType("text/html;charset = UTF - 8");
    try (PrintWriter out = response.getWriter()) {
        //访问与当前请求相关联的会话对象,如果不存在,则创建一个
        HttpSession session = request.getSession();
        //从会话对象中移除名(key)为 msg 的对象
        session.removeAttribute("myMsg");
        //再次从会话对象中取出名(key)为 myMsg 的对象,并强制转换为 String 类型
        String msg = (String) session.getAttribute("myMsg");
        out.print(msg);
        //输出水平线
        out.print("<hr />");
        String sessionId = session.getId();
        out.print(sessionId);
    }
}
```

运行程序,从 index. html 开始。在 index. html 页面输入想说的话"hello",运行结果如图 4.4 所示。

结论:

(1) 三次输出的 ID 相同,说明是同一用户,是同一个 HttpSession 对象。

图 4.4 在会话对象中存入、读取和移除信息

(2) SessionServlet2 中能够输出"hello"，说明 HttpSession 对象的 setAttribute(String name,Object value)方法可以在会话对象中存入数据，HttpSession 对象的 getAttribute (String name)方法可以从会话对象中读取数据。

(3) SessionServlet3 中输出 null，说明 HttpSession 对象的 removeAttribute(String name)方法可以从会话对象中移除数据。

(4) HttpSession 对象可以跨越多个请求，在会话对象被销毁之前，只要信息未被移除，那么存放在会话对象中的数据就可以被同一用户的多个请求所共享。

4.4 浏览器会话与服务器会话的区别

会话过程涉及两个方面：浏览器和服务器，事实上会话也就分为了两个部分——浏览器端的会话和服务器端的会话。在之前的讲解中并没有将这两者刻意区分开，在本小节做一下适当的展开，探讨一下两者的区别。

默认情况下，会话跟踪是基于存储在浏览器内存中的 Cookie 来实现的，也就是说，如果程序没有将 Cookie 保存在客户端的硬盘上，那么一旦关闭浏览器就会造成内存中的 Cookie 消失，从而导致会话跟踪失败。这个就是浏览器端的会话。

如果在服务器端进行会话跟踪，需要在服务器的内存中维护会话对象，在持续不断的会话过程中，这不会造成什么困扰，但是有一种情况例外：假如客户端浏览器已经关闭了，说明这时候浏览器端的会话结束了。但是请注意，服务器端并不知道这个情况，所以服务器端依然需要维护会话对象，直到达到某个约定的时间间隔为止。

这个过程其实一点也不奇怪，以在超市购物为例。进入超市后，超市会为你提供一辆购物车，这个购物车就相当于会话对象，虽然它是超市的资产，但是在你购物的过程中，它归你使用，并且只属于你。超市就相当于服务器。下面分析几个场景。

场景一：你选择了一些商品，并放入了购物车，你将购物车放在附近然后去选择其他商品，这时工作人员是否有权利将购物车推走，将商品放入回收区等待重新上架呢？答案当然是不能。因为你可能正在选购商品，很快就会再次使用这个购物车。

场景二：你选择了一些商品，并放入了购物车，但是你发现出来的匆忙，没带现金和银行卡，没法结账，于是你悄悄地从无购物通道离开了。那么这时工作人员是否有权利将购物车推走，将商品放入回收区等待重新上架呢？答案依然是不能。因为工作人员无法判定你是离开了还是去选择其他商品了。工作人员只能等待，观察一段时间还是没有人用这个购物车，才可以对购物车进行处置。

场景三：你选择了一些商品，并放入了购物车，但是你发现出来的匆忙，没带现金和银行卡，没法结账，但是你没有悄悄从无购物通道离开，而是将车交给工作人员并告诉他，这些商品你不要了，请他帮忙处理一下。工作人员这时候就可以将商品重新上架，并将购物车回收了。

对于场景一、二，超市能做的操作是一样的，那就是被动等待，等待到了一个时间间隔（例如1天，每天打烊后），将购物车处理掉。对于场景三，超市可以立刻进行处理，将商品从购物车取出重新上架或者进行其他约定的操作等。

在Servlet中对会话进行跟踪时也有同超市购物类似的场景，对于用户长时间无操作（对应场景一，用户可能此时在选购其他商品），或者浏览器其实已经关闭的情况（对应场景二，用户已经离开超市），服务器并不能丢弃或者销毁HttpSession对象，只能等待，直至超过预先设置的会话时限。超时之后，会话变为不活动状态，HttpSession对象中存入的对象被移除（解除绑定）。如果想达到场景三类似的效果，那么需要调用相应的方法，例如调用invalidate()方法，将HttpSession对象中存入的对象移除（解除绑定）。

4.5 废弃当前会话对象

废弃会话可以分成被动和主动两种方式。被动方式就是指设置会话超时时间，一旦超出这个时间间隔，服务器就进行解除绑定的操作；另一种就是调用相应的方法主动解除绑定。下面分别进行介绍。

1. 设置会话超时时间

方式一：访问会话对象后，调用它的setMaxInactiveInterval(int interval)方法，其中参数的计时单位为s，若设置会话超时时间为30min，则代码为：

```
HttpSession session = request.getSession();
session.setMaxInactiveInterval(1800);//单位为s
```

方式二：在Web应用程序的配置文件web.xml中设置< session-config >，例如下列代码中设置会话超时时间为30min：

```
<session-config>
    <session-timeout>30</session-timeout>
</session-config>
```

方式三：对于 Tomcat 而言，在 Tomcat 安装目录的 conf 目录下的配置文件 web.xml 中设置<session-config>，例如下列代码中设置会话超时时间为 30min：

```
<session-config>
    <session-timeout>30</session-timeout>
</session-config>
```

其中方式一优先级最高，方式二次之，方式三优先级最低。

对于 GlassFish Server 而言，可以在如下位置找到设置服务器会话超时时间的选项，如图 4.5 所示。

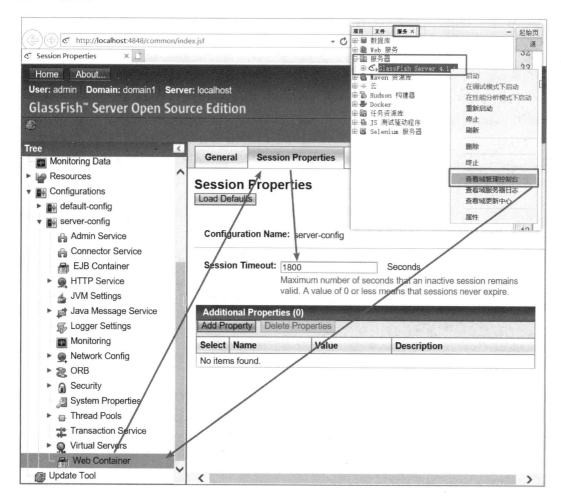

图 4.5 在 GlassFish 中设置会话超时时间

此外,需要注意的是,在 Servlet3.0 中应用程序的部署方式发生了改变,web.xml 不再是必须的,所以按本书中创建的 Web 应用程序,可能找不到 web.xml 文件,如果想要验证<session-config>的设置,新建一个 web.xml 文件即可看到。

设置会话超时,本书就不举例了,感兴趣的读者可以自行操作。

2. 调用 invalidate()方法

下面通过一个简单的示例演示如何通过调用 invalidate()方法废弃会话(解除绑定)。

目标:学习使用 invalidate()方法,废弃会话(解除绑定)。

工程名:eg0403。

用到的文件如表 4.6 所示。

表 4.6 eg0403 用到的文件及文件说明

文件名	说明
SessionServlet.java	手动创建的 Servlet。在这个 Servlet 中首先访问与当前请求相关联的会话,并输出会话 ID;接下来调用 invalidate()方法;最后再次访问与当前请求相关联的会话,并输出会话 ID

编程思路:

比较两次输出的 ID 是否相同,如果不同,则说明两次访问到的会话对象不是同一个,从而证明 invalidate()方法起作用了。

打开 NetBeans IDE,新建一个名为 eg0403 的 Java Web 项目,具体过程略;新建名为 SessionServlet.java 的 Servlet,具体过程略。编写各部分代码如下(仅列出核心代码,大部分自动生成的代码略):

SessionServlet.java 代码片段

```
protected void processRequest(HttpServletRequest request, HttpServletResponse response)
    throws ServletException, IOException {
        response.setContentType("text/html;charset = UTF - 8");
        try (PrintWriter out = response.getWriter()) {
            //访问与当前请求相关联的会话对象,如果不存在,则创建一个
            HttpSession session1 = request.getSession();
            //输出会话 ID
            out.print(session1.getId());
            //调用 invalidate()方法,解除绑定,废弃会话
            session1.invalidate();
            //输出水平线
            out.print("<hr />");
            //再次访问与当前请求相关联的会话对象,如果不存在,则创建一个
            //此处将引用名改为 session2,实际上使用引用名 session1 也可以
            //session1 = request.getSession();
            HttpSession session2 = request.getSession();
            //输出会话 ID
            out.print(session2.getId());
        }
}
```

运行结果如图 4.6 所示，两次输出的会话 ID 不同，证明调用 invalidate()方法后，前一个会话已经被废弃，后一个会话 ID 是重新分配的会话对象的 ID。

图 4.6　invalidate()方法的使用

4.6　利用响应(HttpServletResponse)对象内建方法实现 URL 重写

利用 HttpSession 可以方便地实现会话跟踪，但 HttpSession 默认是使用 Cookie 技术进行会话追踪，如果客户端不接受 Cookie 的话，会话跟踪还能成功吗？

为了验证 Cookie 被阻止时的效果，可以先将浏览器的 Cookie 禁止，再重新运行 eg0401 的例子，看看效果如何。

各个浏览器设置 Cookie 的位置各不相同，处理方式也不同，此处以 Win10 操作系统自带的 IE 11 为例。打开 IE 浏览器，在菜单栏选择“工具”→“Internet 选项”→“隐私”选项卡。单击“高级”按钮，在“高级隐私设置”窗口中可以看到，默认是接受 Cookie 的，选择“阻止”单选框，设置为阻止之后单击“确定”按钮，过程如图 4.7 所示。

运行程序，会发现和之前未阻止 Cookie 时并没有什么不同。再用 Win10 自带的浏览器 Edge，阻止 Cookie 后，运行程序也是如此。看起来是否阻止 Cookie 并不影响程序的运行。那么再换成谷歌浏览器 Chrome。在 Chrome 中阻止 Cookie，如图 4.8 所示。

在浏览器地址栏输入 http://localhost:8080/eg0401/index.html(也可以从 IE 浏览器地址栏复制粘贴)，运行程序。运行效果如图 4.9 所示。

通过上述测试，可以清楚地看到，当 Cookie 被客户端阻止后，使用 HttpSession 进行会话跟踪失败了，因为两次会话 ID 不同，说明服务器认为提交请求的是两个不同的用户。

从这个例子也可以看出，不同浏览器对于 Cookie 的处理是不同的。即使同为 IE 浏览器，在 IE 9 中是可以复现 Chrome 的运行效果的，而 IE 11 中就不行，这个现象很奇怪。Web 程序开发人员无法预见用户到底使用什么版本的浏览器，以及是否阻止 Cookie。对于比较大的网站，例如 Facebook、Twitter，可以对用户做强制性的要求。而对于多数中小站点而言，是没有这个能力去做的(公司内部使用的管理信息系统除外，公司是可以对员工做出强制性要求的)。

那么如果用户使用的是类似于 Chrome 浏览器，并且阻止了 Cookie，该如何实现会话跟踪呢？这就需要采用 URL 重写的方式来实现会话跟踪，它的原理是将 sessionID 作为参数

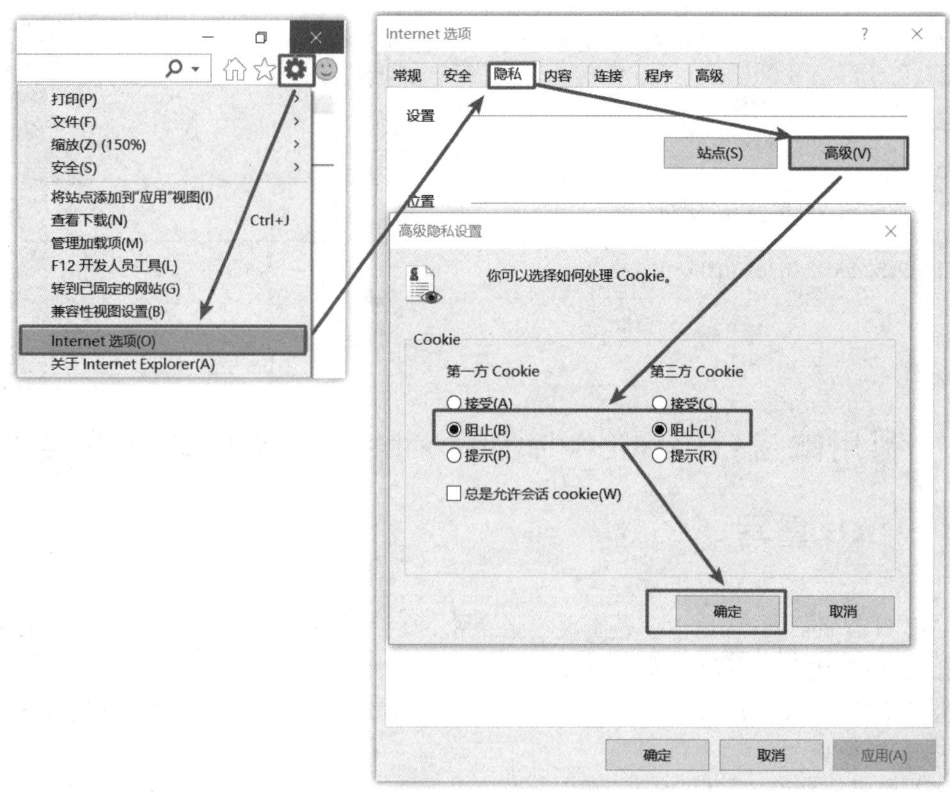

图 4.7 通过隐私设置阻止 Cookie

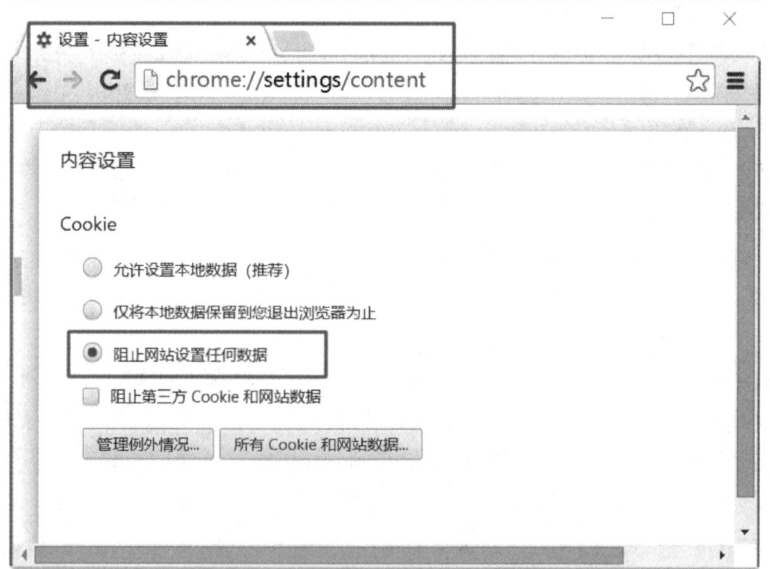

图 4.8 在 Chrome 中阻止 Cookie

图 4.9　在 Chrome 中运行阻止 Cookie 后的 eg0401

附在 URL 后面。当然，如何生成 sessionID，又如何把它附在 URL 后面也不用自己操心，可以利用响应(HttpResponse)对象的内建方法来实现。

通过响应(HttpResponse)对象的 encodeURL(或 encodeRedirectURL)方法可以实现 URL 重写，如表 4.7 所示，这两个方法首先判断浏览器是否支持 Cookie：如果支持，则参数 URL 被原样返回，session ID 将通过 Cookie 来维持；否则返回带有 sessionID 的 URL。

表 4.7　javax.servlet.http.HttpServletResponse 定义的方法

方　法　名	返回值类型	适 用 情 况
encodeURL(String url)	String	通过将会话 ID 包含在指定 URL 中对该 URL 进行编码，如果不需要编码，则返回未更改的 URL。此方法的实现包含可以确定会话 ID 是否需要在 URL 中编码的逻辑。例如，如果浏览器支持 Cookie，或者关闭了会话跟踪，则 URL 编码就不是必须的。对于健壮的会话跟踪，Servlet 发出的所有 URL 都应该通过此方法运行；否则，URL 重写不能用于不支持 Cookie 的浏览器
encodeRedirectURL(String url)	String	对指定 URL 进行编码，以便在 sendRedirect()方法中使用它，如果不需要编码，则返回未更改的 URL。此方法的实现包含可以确定会话 ID 是否需要在 URL 中编码的逻辑。因为进行此确定的规则可能不同于用来确定是否对普通链接进行编码的规则，所以此方法与 encodeURL 方法是分开的。 发送到 HttpServletResponse.sendRedirect()方法的所有 URL 都应该通过此方法运行；否则，URL 重写不能用于不支持 Cookie 的浏览器

下面通过 eg0404 来学习通过响应(HttpResponse)对象内建方法实现 URL 重写。

目标：学习使用响应(HttpResponse)对象内建方法实现 URL 重写。

工程名：eg0404。

eg0404 练习中逻辑结构及大部分代码与 eg0401 相同，所以复制项目 eg0401，设置项目名称为 eg0404。例如在 SessionServlet1 中，只需要将代码

```
out.print("<a href = \"SessionServlet2\">链接到 SessionServlet2 </a>");
```

改为

```
String url = response.encodeURL("SessionServlet2");
out.print("<a href = \"" + url + "\">链接到 SessionServlet2 </a>");
```

运行效果如图 4.10 所示。可以看到在地址栏里面有一个名字为 jsessionid 的参数，它就代表着会话 ID。经过 URL 重写，可以看到，会话跟踪成功。

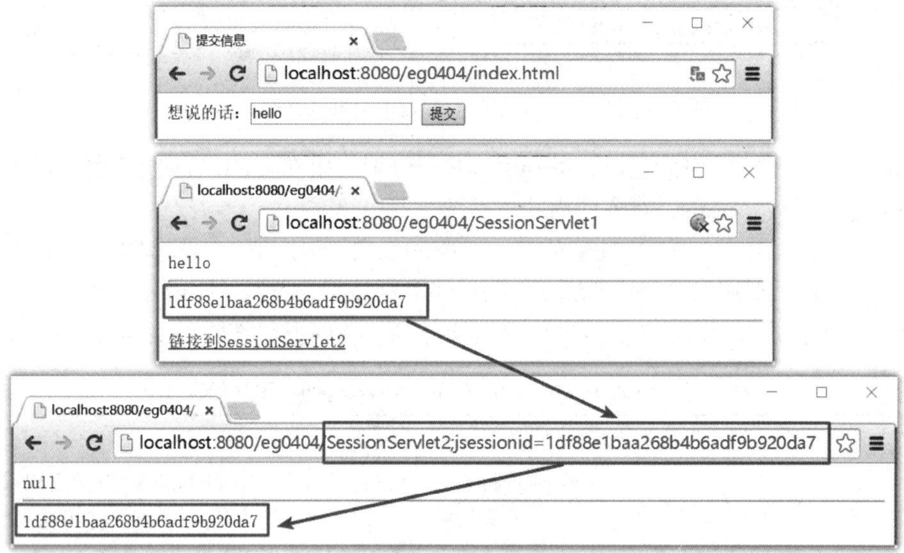

图 4.10 URL 重写之后的会话跟踪

4.7 本章回顾

本章主要介绍了 Servlet 的会话跟踪。简要介绍了什么是会话，为什么需要会话跟踪，常用的会话跟踪技术；讲解了如何获得会话对象；会话接口 HttpSession 及其常用方法，重点介绍了与属性相关的方法；浏览器会话与服务器会话的区别；如何废弃当前会话。最后讲解了如何利用 HttpResponse 对象来实现 URL 重写，如图 4.11 所示。

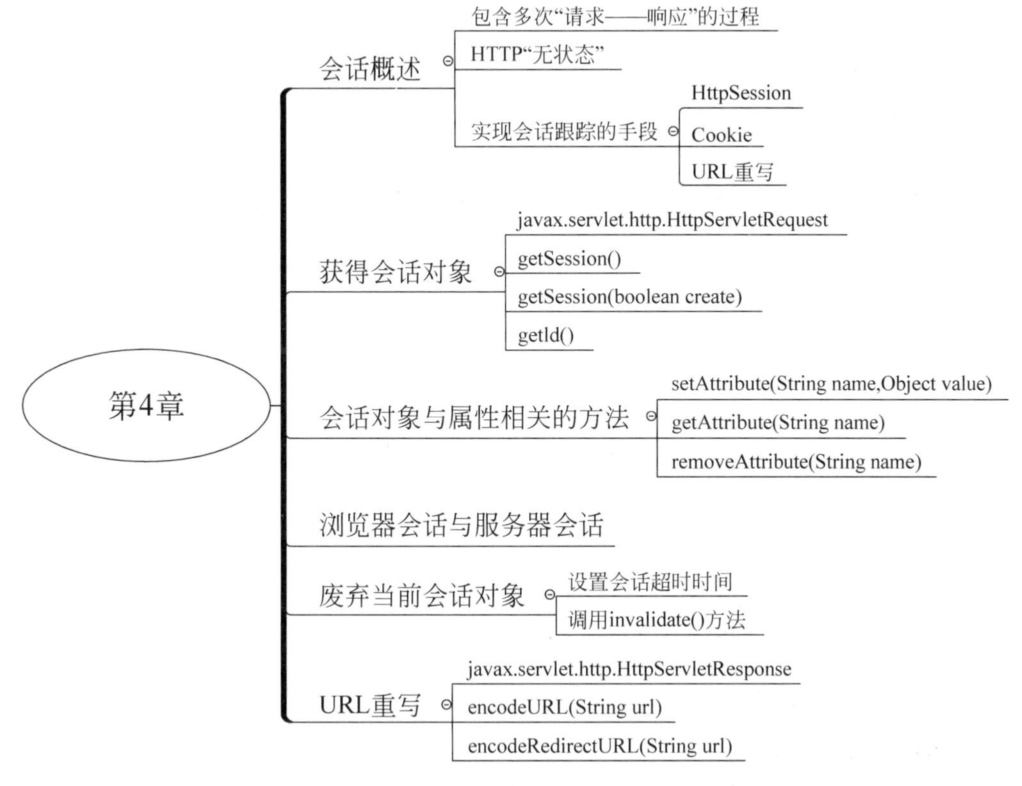

图 4.11 第 4 章内容结构图

4.8 课后习题

1. 常用的会话跟踪技术有哪些？试简要介绍各个技术的基本原理。
2. 如何访问与当前请求相关联的会话对象？这些方法有什么差别？
3. 通过会话对象的什么方法能够在会话中存入、读取和移除信息？
4. 浏览器会话与服务器会话有什么区别？
5. 什么是 URL 重写？如何实现 URL 重写？
6. 尝试使用会话对象实现一个购物车。
7. 尝试使用会话对象，模拟如下效果：当用户登录网站后，访问该网站任何链接，页面上均可显示如下文字“欢迎您，xxx”，xxx 为用户登录名。

第5章 Servlet数据库访问基础

学习目标：

通过本章的学习，你应该：

- 了解 JDBC API 的作用
- 掌握通过 JDBC 访问数据库的步骤
- 掌握使用 NetBeans IDE 管理数据库的方法
- 掌握对数据库进行增、删、改、查的方法
- 熟练掌握使用预编译语句对象对数据库进行增、删、改、查操作

通过前 4 章的学习，已经学习了如下知识：

(1) 通过 HTML，从客户端向服务器提交数据。

(2) 通过 Servlet 提供的 HttpServletRequest 对象，在服务器取得提交的数据并进行处理。

(3) 通过 HttpServletResponse 对象给客户以响应。

(4) 通过 HttpSession 对象对客户进行会话跟踪。

本章将学习如何在 Servlet 中通过 JDBC 访问关系数据库。

5.1 JDBC 连接数据库概述

1. JDBC 简介

Java 数据库连接(Java Database Connectivity，JDBC)是 Java 语言的应用程序接口，用来规范客户端程序对数据库的访问。例如提供查询和更新数据库中数据的方法。通过 JDBC API，可以用统一的形式访问不同的关系数据库产品。JDBC API 已经包含在 JDK 中，导入相应包就可以使用。

JDBC API 中定义了一系列的接口，程序设计人员只需要针对接口进行业务逻辑的编程即可，不必关注接口的实现。接口的实现由数据库厂商提供。使用不同的数据库产品，就需要使用对应厂商提供的接口实现类，这些类通常打包在一起，被称为 JDBC 驱动程序。

JDBC 驱动程序有 4 类，从效率上讲，最高的是"本地协议的纯 Java 驱动程序"，这类驱动程序用 Java 语言编写，通过与数据库建立直接的套接字连接，采用具体厂商的网络协议把 JDBC API 调用转换为直接的网络调用。采用"本地协议的纯 Java 驱动程序"连接数据库的模式如图 5.1 所示。

其他类型的驱动程序当然也有特定用途，本书不详述它们的区别及原理，如感兴趣，请自行搜索相关文章。

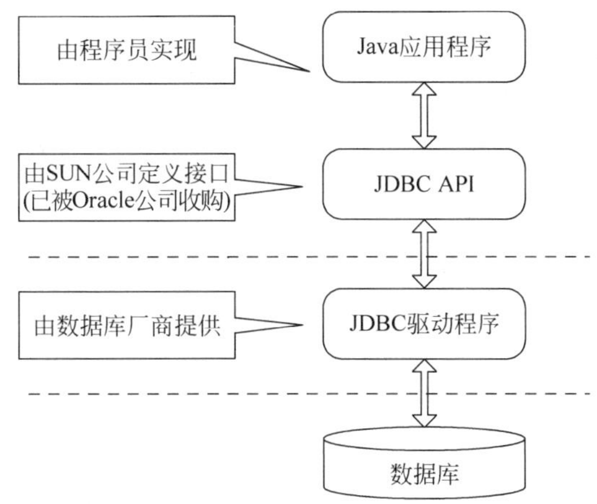

图 5.1　通过“本地协议的纯 Java 驱动程序”连接数据库

2. JDBC 访问数据库的步骤

要想操纵数据库中的数据，首先要连接数据库，那么连接数据库需要知道哪些信息呢？

首先，要知道是什么数据库，以便载入对应厂商的数据库驱动程序。

其次，要知道数据库服务器的位置在哪里，想要访问的是数据库服务器上的哪个数据库，访问该数据库服务器的用户名和密码是什么。

最后，编写 SQL 语句操纵数据库里面的数据。

知道了这些信息，才可以对数据库进行操作。使用 JDBC 对数据库进行一次数据库操作一般需要如下 5 个步骤：

(1) 载入 JDBC 驱动程序。

(2) 通过驱动管理器获得连接对象。

(3) 通过连接对象创建语句对象。

(4) 通过语句对象向服务器发送 SQL 语句，完成业务逻辑。

(5) 依次关闭中间用到的对象(如：结果集对象、语句对象、连接对象)。

5.2　NetBeans IDE 中如何管理数据库

虽然数据库服务器和 NetBeans IDE 是相互独立的，而且一般数据库都有自己的管理工具，但是如果能在 NetBeans IDE 中一并对数据库的操作进行必要的管理，那就更好了，至少不用在多个工具之间来回切换。本节将简单介绍如何在 NetBeans 中管理数据库。

打开 NetBeans IDE。在最常用的“项目”窗口右侧，紧邻“项目”标签的是“文件”标签，接着是“服务”标签，单击“服务”标签切换到“服务”窗口。在服务窗口中可以进行数据库的管理以及服务器的管理等操作。如图 5.2 所示(如果进入“服务”窗口后一开始看不到

sample 数据库和连接,请单击图 5.2 中的“服务器”选项,就会自动生成)。

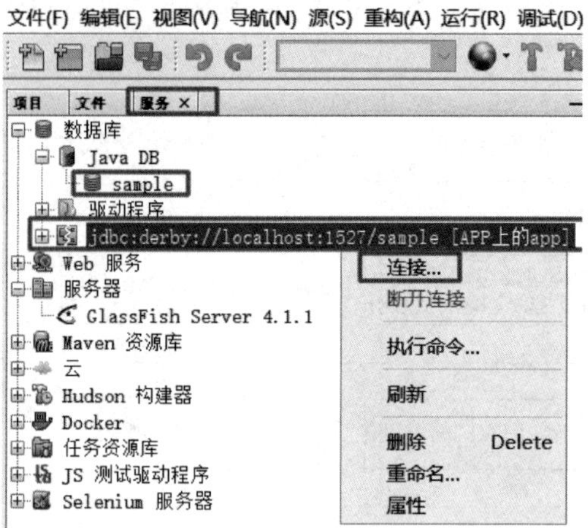

图 5.2 选择服务窗口

从图 5.2 中可以看到,数据库下面有一个 Java DB 子菜单,这个就是之前安装 GlassFish Server 时已经预先安装的数据库服务器,本节将通过对这个数据库产品的操作来学习 Servlet 数据库访问相关的知识。

单击“+号图标”展开 Java DB 菜单,下面是样例数据库 sample。在“驱动程序”菜单选项的下方是连接 sample 数据库的连接,右击该连接,选择“连接”命令。(如果没有显示该连接,可以选择右击 sample,在右键选项中选择“连接”也可以达到目的)。选择“连接”后的效果如图 5.3 所示。

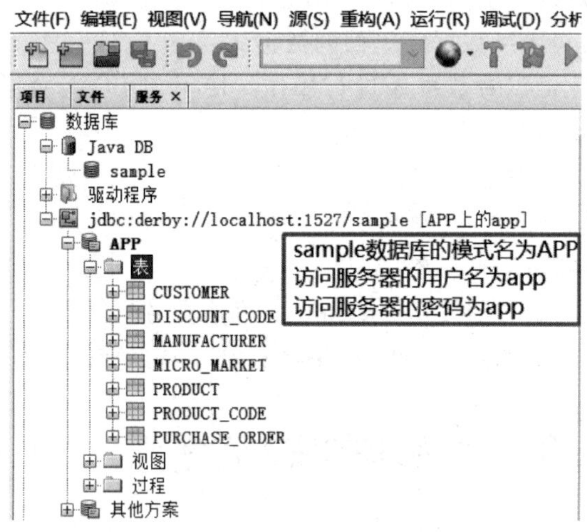

图 5.3 打开数据库连接观察样例表

在连接上右击,选择“属性”命令,弹出窗口如图 5.4 所示。

属性窗口中包含了连接数据库必备的信息,对于初学者来说,复制粘贴这些必备信息相对于照着敲代码,成功率要高得多。通过“数据库 URL”,可以知道数据库服务器在哪里以及连接的是哪个数据库;通过“驱动程序”,能得到驱动程序类的信息;通过“用户”,能知道

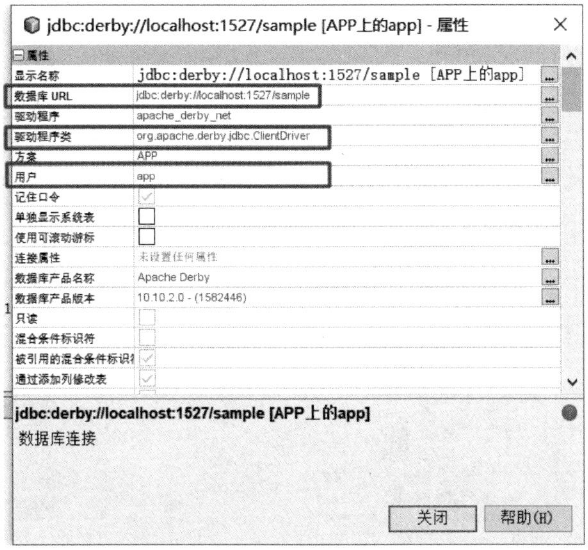

图 5.4　选择连接属性，观察连接的属性中包含的信息

访问数据库的用户名为 app。此外，样例数据库访问密码默认也是 app。把这些信息提取出来以代码形式表示就是下面这样的。

```
String url = "jdbc:derby://localhost:1527/sample";
String driver = "org.apache.derby.jdbc.ClientDriver";
String user = "app";
String password = "app";
```

了解了以上操作，就可以通过编程对数据库进行操作练习了。

5.3　使用 Statement 语句对象进行简单查询操作

下面通过一个简单的示例演示如何在 Servlet 中通过 JDBC，使用 Statement 语句对象进行简单查询操作。

目标：学习在 Servlet 通过 JDBC 对数据库进行简单查询操作。

工程名：eg0501。

已知条件如表 5.1、图 5.5 所示。

表 5.1　eg0501 已知条件

数据库	样例数据库 sample	
表名	customer	（表结构如图 5.5 所示）
欲查询的字段名	customer_id name	数据类型为 INTEGER 数据类型为 VARCHAR(30)

用到的文件如表 5.2 所示。

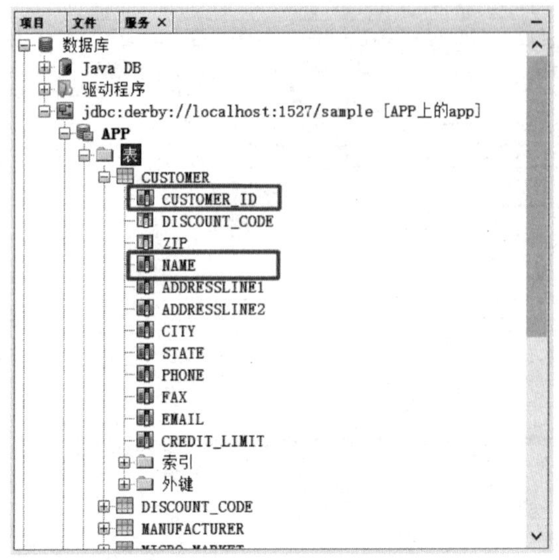

图 5.5 customer 表的结构

表 5.2 eg0501 用到的文件及文件说明

文 件 名	说 明
SimpleQueryServlet.java	手动创建的 Servlet,用来取得 customer 表中的信息,并以网页形式显示

编程思路：

(1) 声明并初始化需要用到的对象。

(2) 声明并初始化连接数据库所需要的参数信息。

(3) 按步骤操作数据库。

打开 NetBeans IDE,新建一个名为 eg0501 的 Java Web 项目,具体过程略。新建名为 SimpleQueryServlet.java 的 Servlet,具体过程略。编写各部分代码如下(仅列出核心代码,大部分自动生成的代码略)。

1. 声明并初始化需要用到的对象

编程过程中需要用到连接对象 Connection、语句对象 Statement、结果集对象 ResultSet。

首先声明连接对象,编写代码如图 5.6 所示。声明过程中会提示红色波浪线,表示代码有问题,将鼠标移动到左侧黄色灯泡处有进一步的提示“找不到符号”。

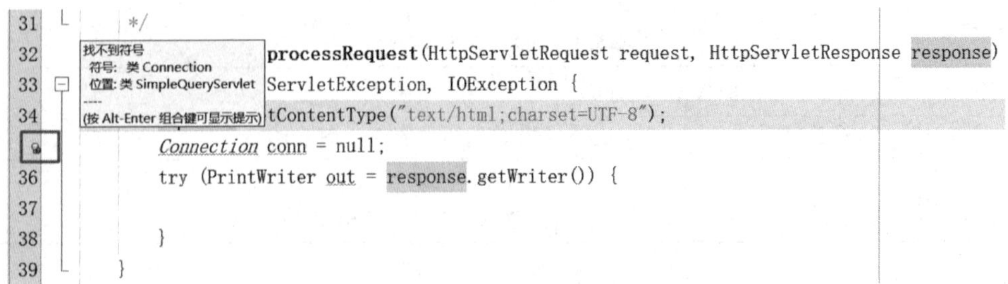

图 5.6 错误提示信息“找不到符号”

单击黄色灯泡处，选择正确的类导入，如图 5.7 所示。

图 5.7　通过错误提示导入 java.sql.Connection 类

还可以使用组合键 Ctrl+Shift+I，自动导入需要的类，如图 5.8 所示。

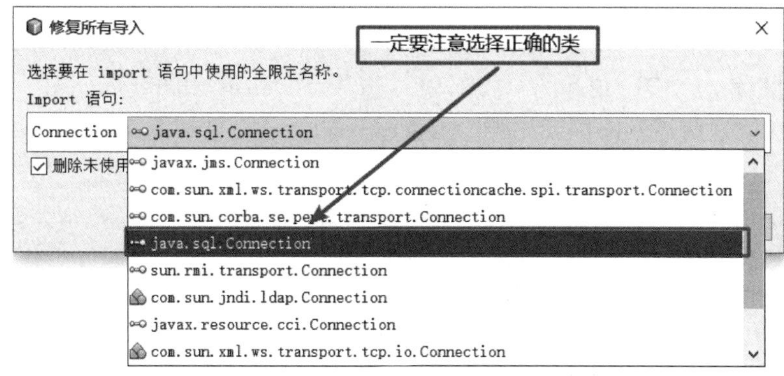

图 5.8　通过 Ctrl+Shift+I 组合键导入 java.sql.Connection 类

还可以在编写代码过程中就选择导入对应的类，推荐此种方式。操作方式是先输入类名或者接口名的前几个字母，然后按快捷键"Ctrl+\"，NetBeans IDE 会给出建议，选择正确的类即可，如图 5.9 所示。

图 5.9　通过"Ctrl+\"提示导入 java.sql.Connection 类

其他类编写过程中出现此类问题也照此办理，注意导入时，一定要选择 java.sql 包中的类，初学时最容易犯的错误就是导入的类不正确。

声明并初始化对象的代码如下：

```
Connection conn = null;
Statement st = null;
ResultSet rs = null;
```

2. 声明并初始化连接数据库所需要的参数信息

这部分信息从图 5.4 所示窗体中得到，代码如下：

```
String url = "jdbc:derby://localhost:1527/sample";
String driver = "org.apache.derby.jdbc.ClientDriver";
String user = "app";
String password = "app";
```

查询数据库的 SQL 语句为“String sql = " select customer_id , name from customer";”完成这两部分操作后，SimpleQueryServlet.java 代码如下：

SimpleQueryServlet.java(part1)

```
package cn.edu.djtu;

import java.io.IOException;
import java.io.PrintWriter;
import java.sql.Connection;
import java.sql.ResultSet;
import java.sql.Statement;
import javax.servlet.ServletException;
import javax.servlet.annotation.WebServlet;
import javax.servlet.http.HttpServlet;
import javax.servlet.http.HttpServletRequest;
import javax.servlet.http.HttpServletResponse;

@WebServlet(name = "SimpleQueryServlet", urlPatterns = {"/SimpleQueryServlet"})
public class SimpleQueryServlet extends HttpServlet {
protected void processRequest(HttpServletRequest request, HttpServletResponse response)
    throws ServletException, IOException {
        response.setContentType("text/html;charset = UTF - 8");

        //连接数据库用到的对象
        Connection conn = null;
        Statement st = null;
        ResultSet rs = null;

        //连接数据库用到的参数信息
        String url = "jdbc:derby://localhost:1527/sample";
        String driver = "org.apache.derby.jdbc.ClientDriver";
```

```
        String user = "app";
        String password = "app";

        //查询数据库的 SQL 语句
        String sql = "select customer_id , name from customer";

        try (PrintWriter out = response.getWriter()) {

    }
}
//doGet()、doPost()等方法略
}
```

3. 接下来按步骤操作数据库

1）载入 JDBC 驱动程序

```
Class.forName(driver );
```

输入 Class. 之后（首字母要大写），NetBeans IDE 会给予语法提示，如果语法提示消失，也可以按快捷键“Ctrl+\”调出语法提示，如图 5.10 所示。

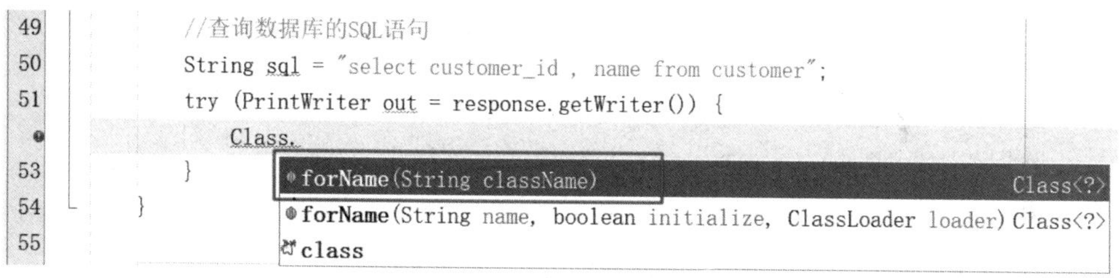

图 5.10　Class.forName(String className)在 IDE 中的提示

将上述代码全部输入完成后，在屏幕上会发现有的代码下面有红色波浪线，表示该代码有问题，将鼠标移动到左侧黄色灯泡图标处有进一步的提示“未报告的异常错误 ClassNotFoundException”，这是因为 Class 的 forName(String className)方法在给出的参数 className 不正确时会抛出异常。提示信息如图 5.11 所示。

```
未报告的异常错误ClassNotFoundException; 必须对其进行捕获或声明以便抛出
将返回值赋给新变量
----
(按 Alt+Enter 组合键可显示提示)
me from customer";
iter()) {
        Class.forName(driver);
    }
}
```

图 5.11　“未报告的异常错误 ClassNotFoundException”提示

此时,单击黄色灯泡图标有可供参考的解决方案,选择“添加 catch 子句”命令,如图 5.12 所示。

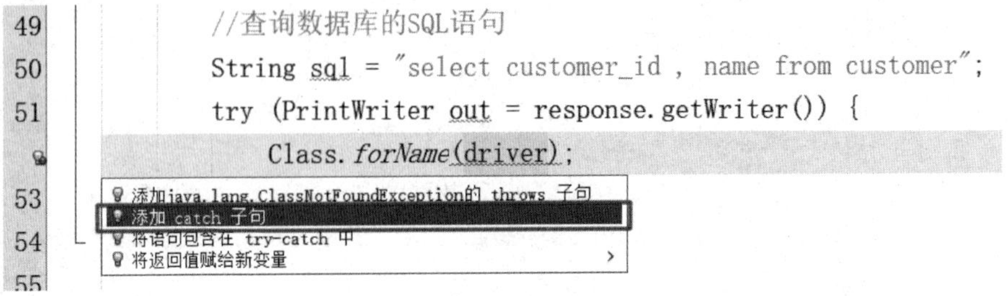

```
        //查询数据库的SQL语句
        String sql = "select customer_id , name from customer";
        try (PrintWriter out = response.getWriter()) {
            Class.forName(driver);
  添加java.lang.ClassNotFoundException的 throws 子句
  添加 catch 子句
  将语句包含在 try-catch 中
  将返回值赋给新变量
```

图 5.12 选择“添加 catch 子句”

2) 通过驱动管理器建立连接

```
conn = DriverManager.getConnection(url, user, password)
```

完成代码后会发现代码下面有红色波浪线,表示代码有问题。原因是 DriverManager 的 getConnection(String url,String user,String password)方法会抛出异常。此时选择与加载驱动时相同的处理方式,根据提示,选择“添加 catch 子句”。NetBeans IDE 此时会有新的提示信息“可以替换为 multicatch”,如图 5.13 所示。

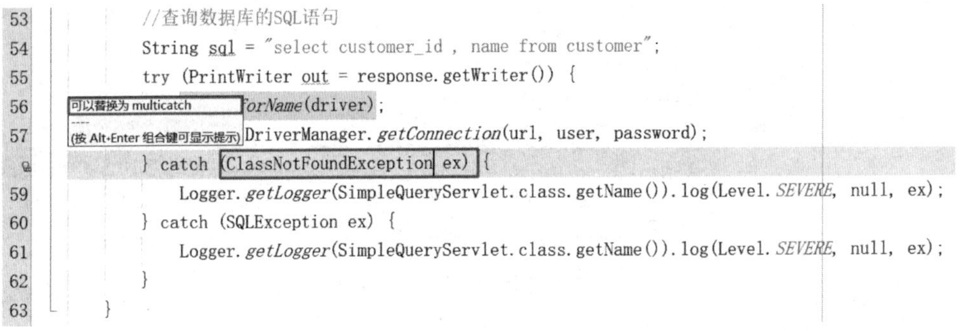

```
        //查询数据库的SQL语句
        String sql = "select customer_id , name from customer";
        try (PrintWriter out = response.getWriter()) {
 可以替换为 multicatch   forName(driver);
 (按 Alt+Enter 组合键可显示提示)  DriverManager.getConnection(url, user, password);
        } catch (ClassNotFoundException ex) {
            Logger.getLogger(SimpleQueryServlet.class.getName()).log(Level.SEVERE, null, ex);
        } catch (SQLException ex) {
            Logger.getLogger(SimpleQueryServlet.class.getName()).log(Level.SEVERE, null, ex);
        }
    }
```

图 5.13 提示信息“可以替换为 multicatch”

按提示操作后,try-catch 部分代码如下:

```
try (PrintWriter out = response.getWriter()) {
    Class.forName(driver);
    conn = DriverManager.getConnection(url, user, password);
} catch (ClassNotFoundException | SQLException ex) {
    Logger.getLogger(SimpleQueryServlet.class.getName()).log(Level.SEVERE, null, ex);
}
```

(1) 通过连接对象创建语句对象。

```
st = conn.createStatement();
```

(2) 通过语句对象向服务器发送 SQL 语句,完成业务逻辑。

因为是查询操作,所以用到执行语句对象的 executeQuery(String sql)方法,返回值类型为 ResultSet。

```
rs = st.executeQuery(sql);
```

然后对结果集进行遍历。在关系数据库中执行查询命令,得到的结果从逻辑上看依然是一张二维表。结果集(ResultSet)对象封装了查询数据库操作的结果集,所以从逻辑结构上也可以认为 ResultSet 是一张二维表。ResultSet 对象包含一个指向当前数据行的游标,游标的初始位置在第一行之前,可以通过 ResultSet 对象的 next 方法向下移动,如表 5.3 所示。

表 5.3 结果集(ResultSet)对象的 next 方法

方法名	返回值类型	适用情况
next()	boolean	移动游标到下一行,如该行存在则返回 true,否则返回 false

定位到一行数据后,如何得到每一列的值呢?

ResultSet 提供了各种 getXxx 方法,它们都以列名或者列索引为参数,返回不同的 Java 数据类型结果。例如: customer_id 字段的数据类型为 INTEGER,可以调用 rs.getInt("customer_id")或者 rs.getInt(1)取得其对应的值。如果仅仅是为了显示结果,那么尽管数据库中 customer_id 的数据类型为 INTEGER,也可以调用 rs.getString("customer_id")或者 rs.getString(1)来取得其对应的值。Name 字段的数据类型为 VARCHAR(30),则需要调用 rs.getString("name")或者 rs.getString(2)来取得其对应的值。

参数中的列索引值指的是该列在 SQL 语句中执行查询操作时,得到的临时表中的列索引,而不是指该列在数据库表中的列索引,在以下 SQL 语句中。

```
String sql = "select customer_id, name from customer";
```

customer_id 列索引值为 1,name 的列索引值为 2,而在 customer 表中:

customer_id 列索引值为 1,name 的列索引值为 4,如果更改 SQL 语句如下:

```
String sql = "select name , customer_id from customer";
```

那么 name 的列索引值为 1,customer_id 列索引值为 2。

访问 ResultSet 的 getXxx 方法时,参数使用列索引,速度较快; 使用列名,便于维护。各有优劣,使用时遵循自己公司的编程规范即可。

因为不知道结果集中有多少行数据,所以遍历时采用 while 循环,完成后代码如下:

```
while(rs.next()){
    out.print(rs.getInt("customer_id"));
    out.print(rs.getString("name"));
}
```

为了让结果看起来更明了,可以考虑用表格格式化输出,调整之后代码如下:

```
out.print("<table border = 1>");
while(rs.next()){
    out.print("<tr>");
        out.print("<td>" + rs.getInt("customer_id") + "</td>");
        out.print("<td>" + rs.getString("name") + "</td>");
    out.print("</tr>");
}
out.print("</table>");
```

(3) 依次关闭中间用到的对象(如:结果集对象、语句对象、连接对象)。

为 try-catch 增加 finally 块,确保关闭对象的操作得到执行。调用相应对象的 close()方法,关闭次序为“先打开的后关闭”,最先创建的对象最后关闭。关闭对象可能抛出 SQLException 异常,所以需要“将语句包含在 try-catch 中”,如图 5.14 所示。

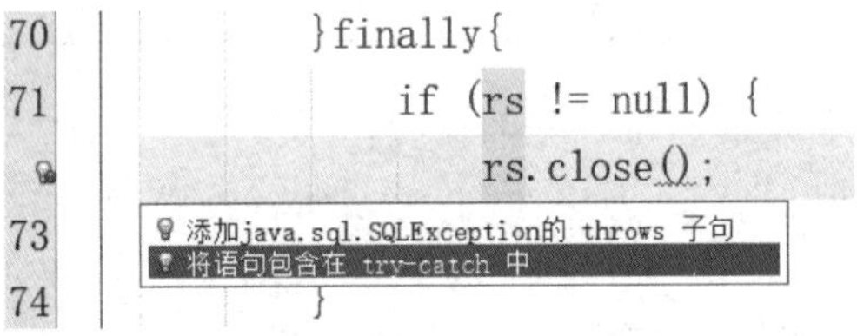

```
}finally{
    if (rs != null) {
        rs.close();
            添加java.sql.SQLException的 throws 子句
            将语句包含在 try-catch 中
    }
```

图 5.14 将关闭结果集对象的语句包含在 try-catch 中

关闭结果集对象。

```
if (rs != null) {
    try {
        rs.close();
    } catch (SQLException ex) {
      Logger.getLogger(SimpleQueryServlet.class.getName()).log(Level.SEVERE, null, ex);
    }
}
```

关闭语句对象。

```
if (st != null) {
    try {
        st.close();
    } catch (SQLException ex) {
        Logger.getLogger(SimpleQueryServlet.class.getName()).log(Level.SEVERE, null, ex);
    }
}
```

关闭连接对象。

```
if (conn != null) {
    try {
        conn.close();
    } catch (SQLException ex) {
        Logger.getLogger(SimpleQueryServlet.class.getName()).log(Level.SEVERE, null, ex);
    }
}
```

全部完成后代码如下：

SimpleQueryServlet.java(all)代码片段

```
package cn.edu.djtu;

import java.io.IOException;
import java.io.PrintWriter;
import java.sql.Connection;
import java.sql.DriverManager;
import java.sql.ResultSet;
import java.sql.SQLException;
import java.sql.Statement;
import java.util.logging.Level;
import java.util.logging.Logger;
import javax.servlet.ServletException;
import javax.servlet.annotation.WebServlet;
import javax.servlet.http.HttpServlet;
import javax.servlet.http.HttpServletRequest;
import javax.servlet.http.HttpServletResponse;

@WebServlet(name = "SimpleQueryServlet", urlPatterns = {"/SimpleQueryServlet"})
public class SimpleQueryServlet extends HttpServlet {

protected void processRequest(HttpServletRequest request, HttpServletResponse response)
  throws ServletException, IOException {
    response.setContentType("text/html;charset=UTF-8");
    //连接数据库用到的对象
    Connection conn = null;
    Statement st = null;
    ResultSet rs = null;

    //连接数据库用到的参数信息
    String url = "jdbc:derby://localhost:1527/sample";
    String driver = "org.apache.derby.jdbc.ClientDriver";
    String user = "app";
    String password = "app";

    //查询数据库的 SQL 语句
    String sql = "select customer_id , name from customer";
    try (PrintWriter out = response.getWriter()) {
```

```
            Class.forName(driver);
            conn = DriverManager.getConnection(url, user, password);
            st = conn.createStatement();
            rs = st.executeQuery(sql);
            out.print("<table>");
              while (rs.next()) {
                out.print("<tr>");
                  out.print("<td>" + rs.getInt("customer_id") + "</td>");
                  out.print("<td>" + rs.getString("name") + "</td>");
                out.print("</tr>");
              }
            out.print("</table>");
          } catch (ClassNotFoundException | SQLException ex) {
            Logger.getLogger(SimpleQueryServlet.class.getName()).log(Level.SEVERE, null, ex);
          }finally{
            if (rs != null) {
              try {
                rs.close();
              } catch (SQLException ex) {
                Logger.getLogger(SimpleQueryServlet.class.getName()).log(Level.SEVERE, null, ex);
              }
            }

            if (st != null) {
              try {
                st.close();
              } catch (SQLException ex) {
                Logger.getLogger(SimpleQueryServlet.class.getName()).log(Level.SEVERE, null, ex);
              }
            }

            if (conn != null) {
              try {
                conn.close();
              } catch (SQLException ex) {
              Logger.getLogger(SimpleQueryServlet.class.getName()).log(Level.SEVERE, null, ex);
              }
            }
          }
        }
            //doGet()、doPost()等方法略
    }
```

代码完成后运行 SimpleQueryServlet。运行结果如图 5.15 所示。

注意：如果使用的服务器不是 GlassFish Server。那么，编写完代码后还需要为项目添加驱动程序。操作步骤如下，有两种选择，一种是"添加库"，另一种是"添加 JAR/文件夹"，一般选择添加库，如图 5.16 所示。

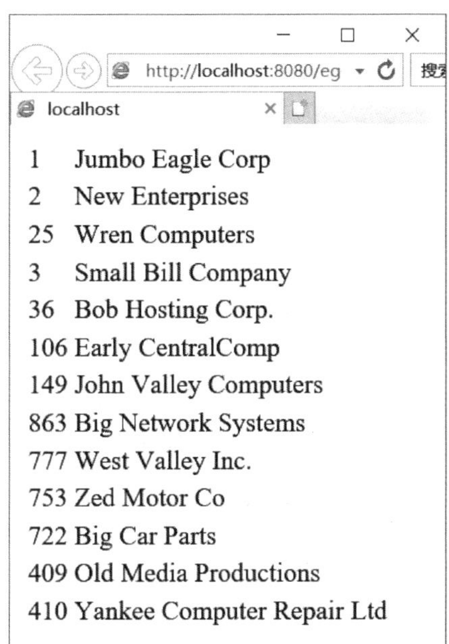

图 5.15　SimpleQueryServlet 运行结果

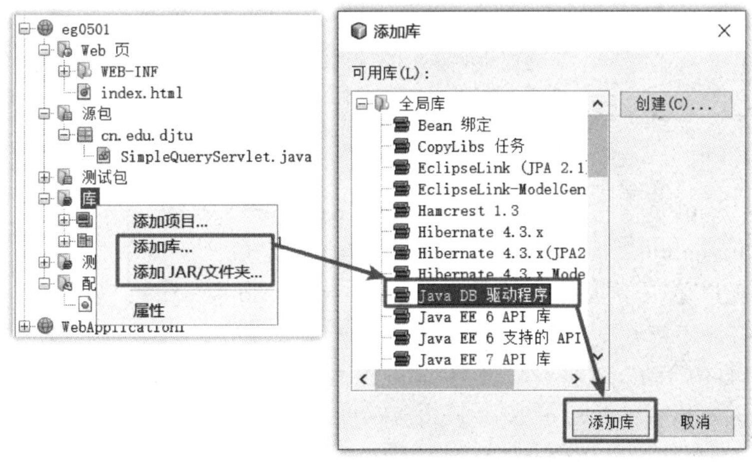

图 5.16　为项目添加库

5.4　使用 Statement 语句对象进行条件查询操作

通过 eg0501 的练习，大家已经对数据库的操作有了初步了解，本节进一步练习条件查询操作。例如：在当当网查询图书，可以按书名检索、按出版社检索等，用户登录其实也涉及条件查询。

对数据库的操作步骤和 eg0501 没有什么大的差别，只是需要多一个页面提交查询条件，并且要将提交的信息“拼”到 SQL 语句中。下面举例说明。

样例 eg0502,演示如何在 Servlet 中使用 Statement 语句对象进行条件查询操作(使用拼接 SQL 语句的方式)。

目标:学习在 Servlet 通过 JDBC 对数据库进行有条件查询操作(使用拼接 SQL 语句的方式)。

工程名:eg0502。

已知条件如表 5.4 所示。

表 5.4 eg0502 已知条件

数据库	样例数据库 sample	
表名	customer	(表结构如图 5.5 所示)
欲查询的字段名	customer_id	数据类型为 INTEGER
	name	数据类型为 VARCHAR(30)
查询条件字段名	discount_code	数据类型为 CHAR(1)

用到的文件如表 5.5 所示。

表 5.5 eg0502 用到的文件及文件说明

文件名	说明
index.html	在创建工程时自动创建的文件,经过改写可用来提交客户 discount_code 信息的页面,页面包含一个表单,表单内有一个文本字段和一个提交按钮
ConditionalQueryServlet.java	手动创建的 Servlet,用来取得用户通过表单提交的查询条件,如果存在满足查询条件的客户则输出客户信息,否则不显示任何信息

编程思路:

(1) index.html 页面将信息提交给 ConditionalQueryServlet。

(2) ConditionalQueryServlet 连接数据库、查询满足条件的客户是否存在。

(3) 如果客户存在,则格式化输出客户的 customer_id 和 name。

(4) 如果客户不存在则不显示信息。

打开 NetBeans IDE,新建一个名为 eg0502 的 Java Web 项目,具体过程略;新建名为 ConditionalQueryServlet.java 的 Servlet,具体过程略。编写各部分代码如下(仅列出核心代码,大部分自动生成的代码略):

index.html

```
<!DOCTYPE html>
<html>
    <head>
        <title>提交信息</title>
        <meta charset = "UTF-8">
        <meta name = "viewport" content = "width = device - width, initial - scale = 1.0">
    </head>
<body>
    <form action = "ConditionalQueryServlet " method = "POST">
        discount_code:<input type = "text" name = "code" value = "" />
```

```
                    <input type = "submit" value = "提交" />
    </form>
</body>
</html>
```

与 SimpleQueryServlet 相比，ConditionalQueryServlet 中首先要取得用户提交的信息，所以增加了这部分语句。

```
//获得用户提交的 code
String code = request.getParameter("code");
```

此外，还需要调整一下 SQL 语句，原来的 SQL 语句如下：

```
String sql = "select customer_id , name from customer";
```

假设客户提交的 discount_code 为 H，那么最终发布给数据库的 SQL 语句应该形如：

```
String sql = "select customer_id , name from customer where discount_code = 'H' " ;
```

注意：因为数据库中 discount_code 字段的数据类型为 CHAR(1)，所以 SQL 语句中需要在 H 两端加上单引号“'”。在实际操作中，H 是从客户提交的表单字段“code”中取到的，取到之后已经赋值给变量code 了。

```
String code = request.getParameter("code");
```

所以，需要将code 这个变量拼接到 SQL 语句中(注意code 变量两边要拼接“ ' ”)，最终发布给数据库的 SQL 语句应该形如：

```
String sql = "select customer_id , name from customer where discount_code = ' " + code + " ' ";
```

调整之后，ConditionalQueryServlet.java 代码如下，与 SimpleQueryServlet 不同的地方以波浪线显示：

ConditionalQueryServlet.java 代码片段

```
package cn.edu.djtu;
//导入包的代码略
@WebServlet(name = "ConditionalQueryServlet", urlPatterns = {"/ConditionalQueryServlet"})
public class SimpleQueryServlet extends HttpServlet {
protected void processRequest(HttpServletRequest request, HttpServletResponse response)
    throws ServletException, IOException {
        response.setContentType("text/html;charset = UTF - 8");

        //获得用户提交的 code
```

```
        String code = request.getParameter("code");

        //连接数据库用到的对象
        Connection conn = null;
        Statement st = null;
        ResultSet rs = null;

        //连接数据库用到的参数信息
        String url = "jdbc:derby://localhost:1527/sample";
        String driver = "org.apache.derby.jdbc.ClientDriver";
        String user = "app";
        String password = "app";

        //查询数据库的SQL语句
        String sql = "select customer_id , name from customer where discount_code = '" +
code + "'";
    try (PrintWriter out = response.getWriter()) {
        Class.forName(driver);
        conn = DriverManager.getConnection(url, user, password);
        st = conn.createStatement();
        rs = st.executeQuery(sql);
        out.print("<table>");
            while (rs.next()) {
                out.print("<tr>");
                    out.print("<td>" + rs.getInt("customer_id") + "</td>");
                    out.print("<td>" + rs.getString("name") + "</td>");
                out.print("</tr>");
            }
        out.print("</table>");
    } catch (ClassNotFoundException | SQLException ex) {
    Logger.getLogger(ConditionalQueryServlet.class.getName()).log(Level.SEVERE, null,
ex);
    }finally{
        if (rs != null) {
            try {
                rs.close();
            } catch (SQLException ex) {
            Logger.getLogger(ConditionalQueryServlet.class.getName()).log(Level.SEVERE,
null, ex);
            }
        }

            if (st != null) {
                try {
                    st.close();
                } catch (SQLException ex) {
```

```
                Logger.getLogger(ConditionalQueryServlet.class.getName()).log(Level.
SEVERE, null, ex);
                }
            }

            if (conn != null) {
                try {
                    conn.close();
                } catch (SQLException ex) {
                  Logger.getLogger(ConditionalQueryServlet.class.getName()).log(Level.
SEVERE, null, ex);
                }
            }
        }
    }
        //doGet()、doPost()等方法略
}
```

代码完成后就可以测试一下程序。先查看一下 discount_code 可以取哪些值。然后在服务里面找到 customer 表，再在 customer 表上右击，选择“查看数据”，可以看到 discount_code 可以取值 N、M、L、H，如图 5.17 所示。

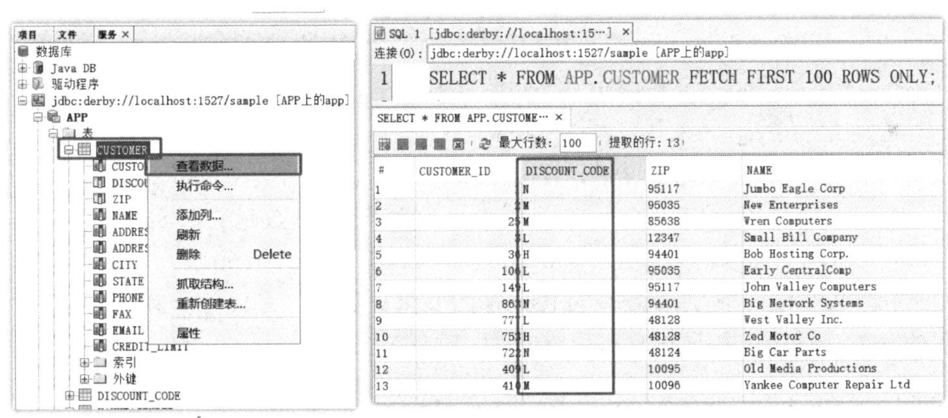

图 5.17 查看 customer 表中的数据

回到项目窗口，运行 index.html，输入 discount_code，例如：N。提交后可看到运行结果，显示了 3 条数据。重新运行 index.html，输入不存在的 discount_code，例如：B。提交可看到运行结果，不显示任何数据，如图 5.18 所示。

拼接 SQL 语句的方式从思路上比较简单，但是随之而来的问题也比较多，当提交的信息较多时，例如在用户注册这样的应用中，会导致 SQL 串过长，如果需要拼接的 CHAR 或者 VARCHAR 过多的话，还得小心不要漏掉变量两边的单引号“ ' ”。

解决条件查询的问题，方案不止一个，下面使用 PreparedStatement（到目前为止的样例中，使用的都是 Statement）来解决它。

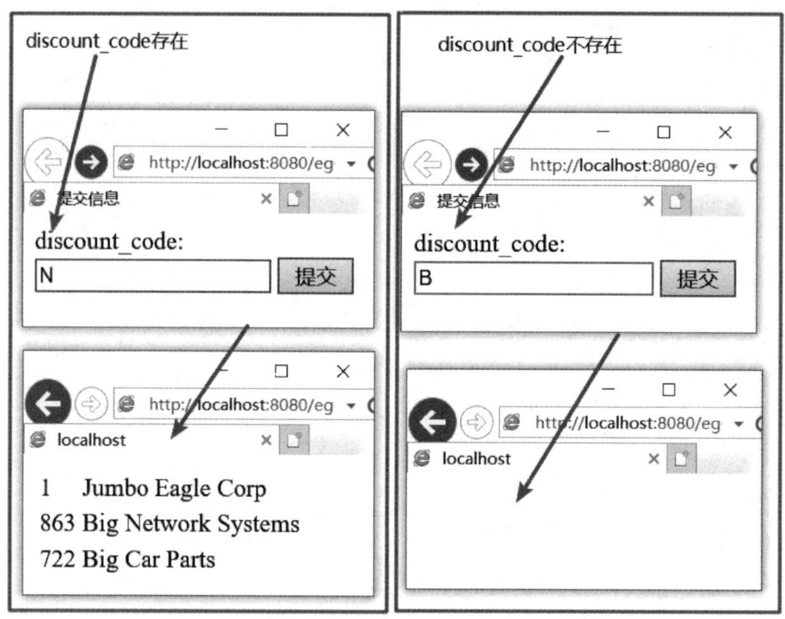

图 5.18 输入 discount_code 时的输出结果对比图

5.5 使用 PreparedStatement 语句对象进行条件查询操作

PreparedStatement 称为预编译的语句对象，它是 Statement 的子接口。通过 Connection 对象的 PrepareStatement(String sql)方法可以获得 PreparedStatement 对象。在多次执行相同 SQL 语句的情况下，PreparedStatement 要比 Statement 高效。

Connection 对象的 PrepareStatement(String sql)方法，参数是一个 String 类型的 SQL 语句(而取得 Statement 时，是调用 Connection 对象的 CreateStatement()方法，是没有参数的)。在这个 SQL 语句中，原本需要拼接进来的变量可以用?来代替。例如：为了生成如下形式的 SQL 语句(其中 H 是动态取得的，需要拼接进去)：

```
String sql = "select customer_id , name from customer where discount_code = 'H'" ;
```

在 eg0502 的 ConditionalQueryServlet 中是这样处理的：

```
String sql = "select customer_id , name from customer where discount_code = '" + code + "'";
```

而通过使用预编译语句(PreparedStatement)对象则不用考虑拼接的问题，可以用如下的 SQL 语句来处理：

```
String sql = "select customer_id , name from customer where discount_code = ? ";
```

? 所代表的动态的值可以通过预编译语句(PreparedStatement)对象的 setXxx 方法来

设置，预编译语句(PreparedStatement)对象常用的 setXxx 方法如表 5.6 所示，具体用到时可查看 Java API。其中 int parameterIndex 值指的是该指定参数在 SQL 语句中的索引值(实质就是看看这个参数在 SQL 语句的所有参数中，从左到右排第几)。索引值的起始值从 1 开始(这一点和数组不同，数组索引值起始下标为 0)。

表 5.6 java.sql.PreparedStatement 定义的方法

方 法 名	返回值类型	适 用 情 况
setBoolean(int parameterIndex, boolean x)	void	用 Java 中 boolean 类型的值设置指定参数的值
setDouble(int parameterIndex, double x)	void	用 Java 中 double 类型的值设置指定参数的值
setFloat(int parameterIndex, float x)	void	用 Java 中 float 类型的值设置指定参数的值
setInt(int parameterIndex, int x)	void	用 Java 中 int 类型的值设置指定参数的值
setLong(int parameterIndex, long x)	void	用 Java 中 long 类型的值设置指定参数的值
setShort(int parameterIndex, short x)	void	用 Java 中 short 类型的值设置指定参数的值
setString(int parameterIndex, String x)	void	用 Java 中 String 类型的值设置指定参数的值

例 1：

数据库中 discount_code 字段的数据类型为 CHAR(1)，客户提交的 code 值已经取到。

```
String code = request.getParameter(code);
```

需要用 String 类型的变量 code 的值来设置，则语句为：

```
String sql  =  "select customer_id , name from customer where discount_code =  ?";
PreparedStatement prst = conn.prepareStatement(sql);
prst.setString(1,code);
```

例 2：

数据库中 customer_id 字段的数据类型为 INTEGER，name 字段数据类型为 VARCHAR(30)，客户提交的 ID 值和 NAME 值已经取到。

```
String cid = request.getParameter("cid");
String cname = request.getParameter("cname");
```

将 cid 的值转换成整型再进行设置，cname 的值不用处理，转换后语句为：

```
String sql  =  "select customer_id , name from customer where customer_id  =  ? and name  = ?";
PreparedStatement prst = conn.prepareStatement(sql);
prst.setInt(1, Integer.parseInt(cid));
prst.setString(2,cname);
```

注意：因为各个数据库产品中，支持的数据类型不尽相同，所以 Java 数据类型与数据库中数据类型的对应关系也并非完全一致，需要根据实际使用的数据库产品对照。

此外，因为 SQL 语句已经预编译过，所以执行读(查询)、写(插入、更新、删除)操作所调用的方法也要调整，读调用 executeQuery 方法，写调用 executeUpdate 方法，都是无参的方法，如表 5.7 所示。

表 5.7 java.sql.PreparedStatement 定义的 executeQuery 方法

方法名	返回值类型	适用情况
executeQuery()	ResultSet	执行查询操作，返回结果集对象
executeUpdate()	int	执行(DML)语句，例如 INSERT、UPDATE、DELETE，返回受影响的结果行数；或者执行 DDL 语句，什么也不返回

下面通过样例 eg0503 来演示如何使用 PreparedStatement 语句对象进行条件查询操作。

目标：学习在 Servlet 通过使用 PreparedStatement 语句对象进行条件查询操作。

工程名：eg0503。

已知条件如表 5.8 所示。

表 5.8 eg0503 已知条件

数据库	样例数据库 sample	
表名	customer	(表结构如图 5.5 所示)
欲查询的字段名	customer_id name	数据类型为 INTEGER 数据类型为 VARCHAR(30)
查询条件字段名	discount_code	数据类型为 CHAR(1)

用到的文件如表 5.9 所示。

表 5.9 eg0503 用到的文件及文件说明

文件名	说明
index.html	为创建工程时自动创建的文件，对其进行简单改写即可。用来提交客户 discount_code 信息的页面，包含一个表单，表单内有一个文本字段和一个提交按钮
ConditionalQueryServlet.java	手动创建的 Servlet，用来取得用户通过表单提交的查询条件，如果满足查询条件的客户存在则输出客户信息，否则不显示任何信息

编程思路：

(1) index.html 页面提交信息给 ConditionalQueryServlet。

(2) ConditionalQueryServlet 连接数据库、查询满足条件的客户是否存在。

(3) 如果客户存在，则格式化输出客户的 customer_id 和 name。

(4) 如果客户不存在则不显示信息。

打开 NetBeans IDE，新建一个名为 eg0503 的 Java Web 项目，具体过程略。新建名为 ConditionalQueryServlet.java 的 Servlet，具体过程略。编写各部分代码如下(仅列出核心代码，大部分自动生成的代码略)：

index.html

```
<!DOCTYPE html>
<html>
    <head>
        <title>提交信息</title>
        <meta charset="UTF-8">
        <meta name="viewport" content="width=device-width, initial-scale=1.0">
    </head>
<body>
    <form action="ConditionalQueryServlet " method="POST">
        discount_code:<input type="text" name="code" value="" />
                        <input type="submit" value="提交" />
    </form>
</body>
</html>
```

与上一版 ConditionalQueryServlet.java 相比，大部分代码相同，代码不同的地方如下面代码中波浪线部分代码所示。

ConditionalQueryServlet.java 代码片段

```
package cn.edu.djtu;
//导入包的代码略
@WebServlet(name = "ConditionalQueryServlet", urlPatterns = {"/ConditionalQueryServlet"})
public class SimpleQueryServlet extends HttpServlet {
protected void processRequest(HttpServletRequest request, HttpServletResponse response)
  throws ServletException, IOException {
    response.setContentType("text/html;charset=UTF-8");

    //获得用户提交的 code
    String code = request.getParameter("code");

    //连接数据库用到的对象
    Connection conn = null;
    PreparedStatement prst = null;
    ResultSet rs = null;

    //连接数据库用到的参数信息
    String url = "jdbc:derby://localhost:1527/sample";
    String driver = "org.apache.derby.jdbc.ClientDriver";
    String user = "app";
    String password = "app";

    //查询数据库的 SQL 语句
    String sql = "select customer_id , name from customer where discount_code = ?";
    try (PrintWriter out = response.getWriter()) {
      Class.forName(driver);
      conn = DriverManager.getConnection(url, user, password);
```

```
        prst = conn.prepareStatement(sql);
        prst.setString(1,code);
        rs = prst.executeQuery();
        out.print("<table>");
          while (rs.next()) {
            out.print("<tr>");
              out.print("<td>" + rs.getInt("customer_id") + "</td>");
              out.print("<td>" + rs.getString("name") + "</td>");
            out.print("</tr>");
          }
        out.print("</table>");
    } catch (ClassNotFoundException | SQLException ex) {
      Logger.getLogger(ConditionalQueryServlet.class.getName()).log(Level.SEVERE, null,
ex);
    }finally{
        if (rs != null) {
          try {
            rs.close();
          } catch (SQLException ex) {
            Logger.getLogger(ConditionalQueryServlet.class.getName()).log(Level.SEVERE,
null, ex);
          }
        }

        if (prst != null) {
          try {
            prst.close();
          } catch (SQLException ex) {
            Logger.getLogger(ConditionalQueryServlet.class.getName()).log(Level.SEVERE,
null, ex);
          }
        }

        if (conn != null) {
          try {
            conn.close();
          } catch (SQLException ex) {
            Logger.getLogger(ConditionalQueryServlet.class.getName()).log(Level.SEVERE,
null, ex);
          }
        }
      }
  }
    //doGet()、doPost()等方法略
}
```

运行结果略。

除了从效率上考虑，从安全性上考虑，使用 PreparedStatement 还可以防止“SQL 注入”，至于什么是“SQL 注入”，请自行搜索。

一起做个实验：如果在 eg0502 的 index.html 页面上输入括号内的文字('青椒麦客'or '1'='1')，很显然这并不是一个合理的 discount_code，但是运行时会发现将显示出全部的数据，如图 5.19 所示。

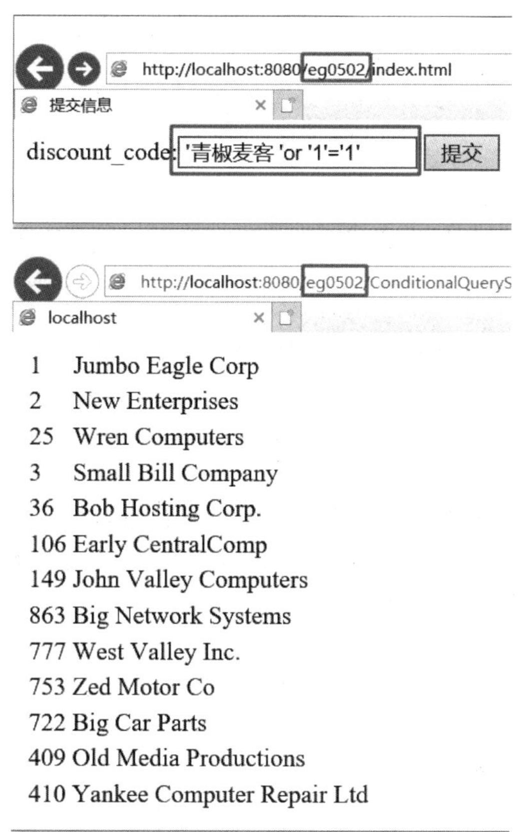

图 5.19　SQL 注入运行示例

这是因为使用拼接 SQL 的方式实际的 SQL 语句变成了：

```
select customer_id , name from customer where discount_code = '青椒麦客' or '1' = '1'
```

虽然没有输入任何合理的 code 的值，但是 where 子句中 or 后面的条件'1'='1'是恒成立的，导致 where 条件中无论 discount_code 取什么值，条件都为真，永远成立。可想而知，如果这是用户登录模块的话，恶意用户可能会绕过用户名和密码的验证而进入系统。

所以，如果在 SQL 语句中用到了动态参数的话，请优先考虑使用 PreparedStatement。

5.6　对数据库进行插入、更新和删除操作的案例准备

1. 对数据库进行插入、更新和删除操作的若干问题

对数据库进行查询，相当于"读"数据库，而对数据库进行插入、更新和删除操作则相当

于“写”数据库。“读”数据库不会对数据库状态造成影响,而“写”数据库则会影响数据库的状态。相应地,在数据库进行“写”操作涉及的问题比“读操作”要多一些,以插入操作为例,就至少存在两个问题。

问题1:“Java中的数据类型与数据库中的数据类型可能不匹配”。

不管用户提交什么信息,通过请求(HttpServletRequest)对象的getParameter(String name)方法返回的都是String类型,而数据库中的数据类型可能是多种多样的,这就需要做数据类型的转换。如果数据类型不匹配,将会抛出SQLException导致插入数据失败。

问题2:“用户输入的数据不合理”。

即使数据类型转换是对的,也无法保证用户提交的信息都是合理的。例如:数据库里面的NAME字段数据类型是VARCHAR(30)。然而由于用户误操作,输入的NAME长度为100;或者需要输入数值型数据的地方输入了字符,这些操作都会导致插入数据失败。

解决问题1,程序员做好数据类型的转换就可以了,比较容易解决。解决问题2要麻烦一些,因为用户的行为是不可控的。而且用户有时也是无心之举,并不是故意为之。问题2的解决一般需要在提交数据时对用户提交的数据进行“前端”校验,这通常使用JavaScript来解决。

为了专心学习数据库的操作,在“写”操作的样例中,暂时假设输入的所有数据都是合理的,并且尽可能简化业务逻辑,以使操作过程更容易理解。在这个简化的逻辑中,将各个操作尽可能独立开来,并不具有实用价值,只是为了学习基本语法知识。

2. 案例用到的表结构及数据

在sample数据库中有一张表叫作DISCOUNT_CODE,表中有两个字段,一个代表折扣码,一个代表折扣率,如表5.10所示。

表5.10 DISCOUNT_CODE表的结构

字段名	数据类型	约束
discount_code	CHAR(1)	PRIMARY KEY
rate	DECIMAL(4,2)	

discount_code表在sample数据库中已经存在,并且已经有数据了。表中的样例数据如表5.11所示。

表5.11 discount_code表中的样例数据

DISCOUNT_CODE	RATE
H	16.00
M	11.00
L	7.00
N	0.00

3. 案例分析与设计

用户访问index.html,在index.html页面可以选择4种不同的操作:

(1) 查询全部 discound_code 和 rate 信息。

(2) 插入 discound_code 和 rate 信息。

(3) 更新 discount_code 和 rate 信息。

(4) 删除 discound_code 和 rate 信息。

整个案例的顺序图如图 5.20 所示。

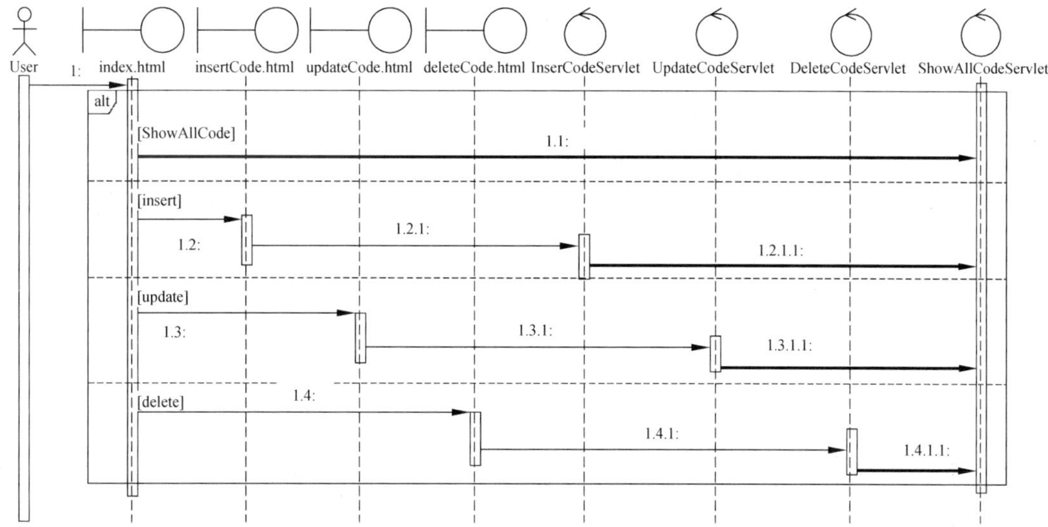

图 5.20 “写”数据库案例顺序图

工程名：eg0504。

用到的文件如表 5.12 所示。

表 5.12 eg0504 用到的文件

文件名	说明
index. html	为创建工程时自动创建的文件。用作导航页面，有 4 个超级链接，分别链接至 ShowAllPersonServlet、InsertCode. html、UpdateCode. html、DeleteCode. html
insertCode. html	通过一个表单将需要插入的新的 discount_code 和 rate 信息提交给 InsertCodeServle
updateCode. html	通过一个表单将需要更新的 discount_code 和 rate 信息提交给 UpdateCodeServle
deleteCode. html	通过一个表单将需要删除的 discount_code 信息提交给 DeleteCodeServle
ShowAllCodeServlet. java	手动创建的 Servlet，显示 discount_code 表的全部数据
InsertCodeServlet. java	手动创建的 Servlet，取得客户提交的 discount_code 和 rate 信息。如果该 discount_code 已经存在，则提示“该 discount_code 已存在!”；否则将 discount_code 和 rate 信息插入到 discount_code 表中。然后将请求转发至 ShowAllCodeServlet 观察操作结果
UpdateCodeServlet. java	手动创建的 Servlet，取得客户提交的 discount_code 和 rate 信息。如果该 discount_code 不存在，则提示“该 discount_code 不存在!”；否则将该 discount_code 的 rate 信息更新。然后将请求转发至 ShowAllCodeServlet 观察操作结果

续表

文　件　名	说　　明
DeleteCodeServlet. java	手动创建的 Servlet,取得客户提交的 discount_code 信息。如果该 discount_code 不存在,则提示“该 discount_code 不存在!”;否则将该 discount_code 对应的记录删除。然后将请求转发至 ShowAllCodeServlet 观察操作结果

4. 案例中导航、查询功能的实现

打开 NetBeans IDE,新建一个名为 eg0504 的 Java Web 项目,具体过程略;分别新建名为 insertCode. html、updateCode. html、deleteCode. html、ShowAllCodeServlet. java、InsertCodeServlet. java、UpdateCodeServlet. java、DeleteCodeServlet. java 的文档(index. html 已经自动创建完毕),具体过程略。

查询操作,可以参照 5. 5 节样例完成,index. html 和 ShowAllCodeServlet. java 文档代码如下:

index. html

```
<!DOCTYPE html>
<html>
  <head>
    <title>首页</title>
    <meta charset = "UTF - 8">
    <meta name = "viewport" content = "width = device - width, initial - scale = 1.0">
  </head>
<body>
    <a href = "ShowAllCodeServlet">显示全部折扣码</a>
    <a href = "insertCode.html">新增折扣码</a>
    <a href = "updateCode.html">更新折扣率</a>
    <a href = "deleteCode.html">删除折扣码</a>
</body>
</html>
```

ShowAllCodeServlet. java 代码片段

```
package cn.edu.djtu;
//导入包的代码略
@ WebServlet ( name = " ConditionalQueryServlet ", urlPatterns = { "/
ConditionalQueryServlet"})
public class ShowAllCodeServlet extends HttpServlet {
protected void processRequest(HttpServletRequest request, HttpServletResponse response)
  throws ServletException, IOException {
    response.setContentType("text/html;charset = UTF - 8");

    //连接数据库用到的对象
    Connection conn = null;
    PreparedStatement prst = null;
```

```
ResultSet rs = null;

//连接数据库用到的参数信息
String url = "jdbc:derby://localhost:1527/sample";
String driver = "org.apache.derby.jdbc.ClientDriver";
String user = "app";
String password = "app";

//查询数据库的SQL语句
String sql = "select discount_code , rate from discount_code";
try (PrintWriter out = response.getWriter()) {
  Class.forName(driver);
  conn = DriverManager.getConnection(url, user, password);
  prst = conn.prepareStatement(sql);
  rs = prst.executeQuery();
  out.print("<table>");
    while (rs.next()) {
      out.print("<tr>");
        out.print("<td>" + rs.getString("discount_code") + "</td>");
        out.print("<td>" + rs.getDouble("rate") + "</td>");
      out.print("</tr>");
    }
  out.print("</table>");
} catch (ClassNotFoundException | SQLException ex) {
  Logger.getLogger(ShowAllCodeServlet.class.getName()).log(Level.SEVERE, null, ex);
}finally{
  if (rs != null) {
    try {
      rs.close();
    } catch (SQLException ex) {
    Logger.getLogger(ShowAllCodeServlet.class.getName()).log(Level.SEVERE, null, ex);
  }
}

if (prst != null) {
  try {
    prst.close();
  } catch (SQLException ex) {
  Logger.getLogger(ShowAllCodeServlet.class.getName()).log(Level.SEVERE, null, ex);
  }
}

if (conn != null) {
  try {
    conn.close();
  } catch (SQLException ex) {
  Logger.getLogger(ShowAllCodeServlet.class.getName()).log(Level.SEVERE, null, ex);
  }
}
```

```
    }
  }
      //doGet()、doPost()等方法略
  }
```

运行效果如图 5.21 所示。

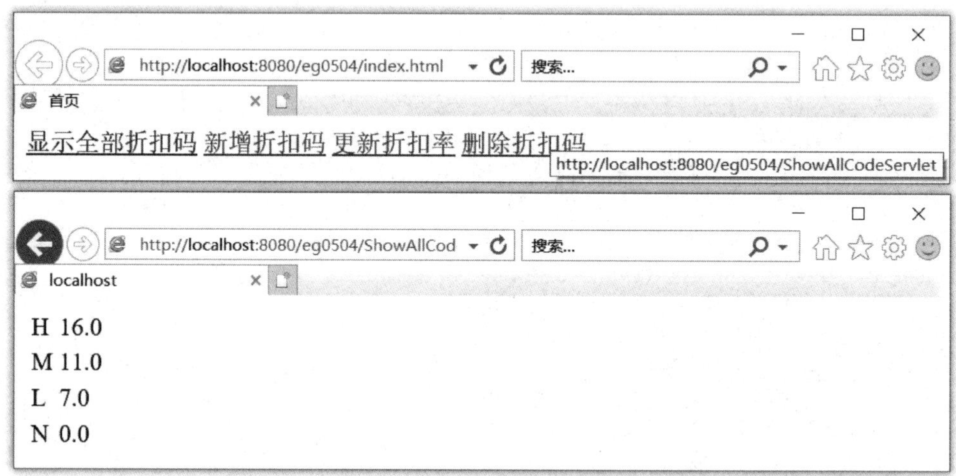

图 5.21　选择查看全部折扣码运行效果

5.7 使用 PreparedStatement 语句对象进行插入操作

首先编写 insertCode.html 文档的代码。

insertCode.html

```
<!DOCTYPE html>
<html>
  <head>
    <title>新增折扣码</title>
    <meta charset = "UTF - 8">
    <meta name = "viewport" content = "width = device - width, initial - scale = 1.0">
  </head>
  <body>
      <form action = "InsertCodeServlet" method = "post">
          <p>discount_code:<input type = "text" name = "discount_code"/></p>
          <p>rate:<input type = "text" name = "rate"/></p>
          <p><input type = "submit" value = "新增"/></p>
      </form>
    </body>
</html>
```

执行插入操作与有条件查询操作差别并不大，主要的差别是两个地方，一个是 SQL 语句不同，另一个是 PreparedStatement 对象执行的方法不同，读操作调用 executeQuery()方法，写操作调用 executeUpdate()方法，此处只给出 InsertCodeServlet 相关核心代码，大部分自动生成的代码略。

InsertCodeServlet 程序活动图如图 5.22 所示。

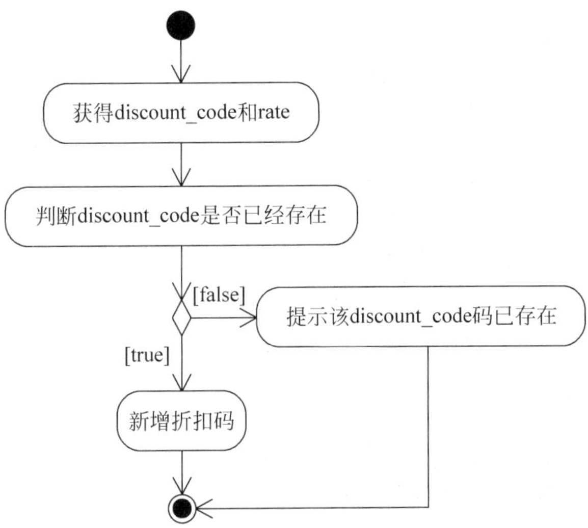

图 5.22　InsertCodeServlet 活动图

InsertCodeServlet.java 代码片段

```
package cn.edu.djtu;
//导入包的代码略
@WebServlet(name = "InsertCodeServle", urlPatterns = {"/InsertCodeServle"})
public class InsertCodeServle extends HttpServlet {
protected void processRequest(HttpServletRequest request, HttpServletResponse response)
  throws ServletException, IOException {
    response.setContentType("text/html;charset = UTF - 8");
    //获得输入的 discount_code 和 rate
    String discount_code = request.getParameter("discount_code");
    String rate = request.getParameter("rate");

    //连接数据库用到的对象
    Connection conn = null;
    PreparedStatement prst = null;
    ResultSet rs = null;

    //连接数据库用到的参数信息
    String url = "jdbc:derby://localhost:1527/sample";
    String driver = "org.apache.derby.jdbc.ClientDriver";
    String user = "app";
    String password = "app";
```

```
        //查询数据库的 SQL 语句
        String sql1 = "select discount_code , rate from discount_code where discount_code = ?";
        //插入新的折扣码的 SQL 语句
        String sql2 = "insert into discount_code values(?,?)";

        try (PrintWriter out = response.getWriter()) {
          Class.forName(driver);
          conn = DriverManager.getConnection(url, user, password);
          prst = conn.prepareStatement(sql1);
          prst.setString(1, discount_code);
          rs = prst.executeQuery();
          if (rs.next()) {
              out.print("该折扣码已存在");
          } else {
              prst = conn.prepareStatement(sql2);
              prst.setString(1, discount_code);
              prst.setDouble(2, Double.parseDouble(rate));
              prst.executeUpdate();
              request.getRequestDispatcher("ShowAllCodeServlet").forward(request, response);
          }
        } catch (ClassNotFoundException | SQLException ex) {
          Logger.getLogger(InsertCodeServle.class.getName()).log(Level.SEVERE, null, ex);
        }finally{
          if (rs != null) {
            try {
              rs.close();
            } catch (SQLException ex) {
            Logger.getLogger(InsertCodeServle.class.getName()).log(Level.SEVERE, null, ex);
            }
          }

          if (prst != null) {
            try {
              prst.close();
            } catch (SQLException ex) {
            Logger.getLogger(InsertCodeServle.class.getName()).log(Level.SEVERE, null, ex);
            }
          }

          if (conn != null) {
            try {
              conn.close();
            } catch (SQLException ex) {
            Logger.getLogger(InsertCodeServle.class.getName()).log(Level.SEVERE, null, ex);
            }
          }
        }
      }
        //doGet()、doPost()等方法略
  }
```

向 discount_code 表中插入已存在的折扣码时，运行效果如图 5.23 所示。

http://localhost:8080/eg0504/index.html

首页

显示全部折扣码 新增折扣码 更新折扣率 删除折扣码

http://localhost:8080/eg0504/insertCode.h

新增折扣码

该折扣码已存在

discount_code: N

rate: 10

新增

http://localhost:8080/eg0504/InsertCodeS

localhost

该折扣码已存在

图 5.23 插入已存在的折扣码时的运行效果

如果插入的数据是尚未存在的折扣码，那么运行效果如图 5.24 所示。

http://localhost:8080/eg0504/insertCode.h

新增折扣码

discount_code: A

rate: 10

新增

http://localhost:8080/eg0504/InsertCodeS

localhost

H 16.0
M 11.0
L 7.0
N 0.0
A 10.0

图 5.24 插入尚未存在的折扣码时的运行效果

5.8 使用 PreparedStatement 语句对象进行更新操作

首先编写 updateCode. html 文档的代码,可以复制粘贴 insertCode. html 代码再修改。

updateCode. html

```
<!DOCTYPE html>
<html>
<head>
<title>更新折扣率</title>
<meta charset = "UTF-8">
<meta name = "viewport" content = "width = device-width, initial-scale = 1.0">
</head>
<body>
<form action = "UpdateCodeServlet" method = "post">
<p>discount_code:<input type = "text" name = "discount_code"/></p>
<p>rate:<input type = "text" name = "rate"/></p>
<p><input type = "submit" value = "更新折扣率"/></p>
</form>
</body>
</html>
```

执行更新操作与执行插入操作基本一致,所以可以复制 InsertCodeServlet 代码再进行改写,大部分自动生成的代码略。UpdateCodeServlet 程序活动图如图 5.25 所示。

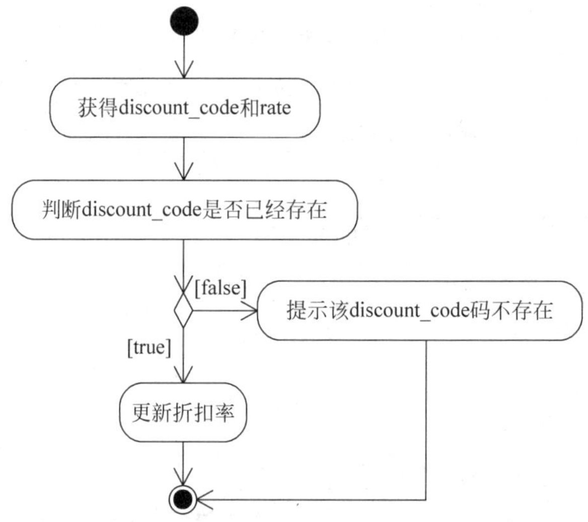

图 5.25 UpdateCodeServlet 活动图

UpdateCodeServlet. java 代码片段

```
package cn.edu.djtu;
//导入包的代码略
```

```
@WebServlet(name = " UpdateCodeServlet ", urlPatterns = {"/ UpdateCodeServlet "})
public class UpdateCodeServlet extends HttpServlet {
protected void processRequest(HttpServletRequest request, HttpServletResponse response)
  throws ServletException, IOException {
    response.setContentType("text/html;charset = UTF - 8");
    //获得输入的 discount_code 和 rate
    String discount_code = request.getParameter("discount_code");
    String rate = request.getParameter("rate");

    //连接数据库用到的对象
    Connection conn = null;
    PreparedStatement prst = null;
    ResultSet rs = null;

    //连接数据库用到的参数信息
    String url = "jdbc:derby://localhost:1527/sample";
    String driver = "org.apache.derby.jdbc.ClientDriver";
    String user = "app";
    String password = "app";

    //查询数据库的 SQL 语句
    String sql1 = "select discount_code , rate from discount_code where discount_code = ?";
    //更新折扣码的 SQL 语句
    String sql2 = "update discount_code set rate = ? where discount_code = ?";

    try (PrintWriter out = response.getWriter()) {
      Class.forName(driver);
      conn = DriverManager.getConnection(url, user, password);
      prst = conn.prepareStatement(sql1);
      prst.setString(1, discount_code);
      rs = prst.executeQuery();
      if (rs.next()) {
          prst = conn.prepareStatement(sql2);
          prst.setDouble(1, Double.parseDouble(rate));
          prst.setString(2, discount_code);
          prst.executeUpdate();
          request.getRequestDispatcher("ShowAllCodeServlet").forward(request, response);
        } else {
            out.print("该折扣码不存在");
        }
      } catch (ClassNotFoundException | SQLException ex) {
        Logger.getLogger(UpdateCodeServlet.class.getName()).log(Level.SEVERE, null, ex);
      }finally{
        if (rs != null) {
          try {
            rs.close();
          } catch (SQLException ex) {
```

```
            Logger.getLogger(UpdateCodeServlet.class.getName()).log(Level.SEVERE, null, ex);
          }
        }

        if (prst != null) {
          try {
            prst.close();
          } catch (SQLException ex) {
          Logger.getLogger(UpdateCodeServlet.class.getName()).log(Level.SEVERE, null, ex);
          }
        }

        if (conn != null) {
          try {
            conn.close();
          } catch (SQLException ex) {
          Logger.getLogger(UpdateCodeServlet.class.getName()).log(Level.SEVERE, null, ex);
          }
        }
      }
    }
    //doGet()、doPost()等方法略
    }
```

如果折扣码存在，更新折扣率时运行效果如图 5.26 所示。

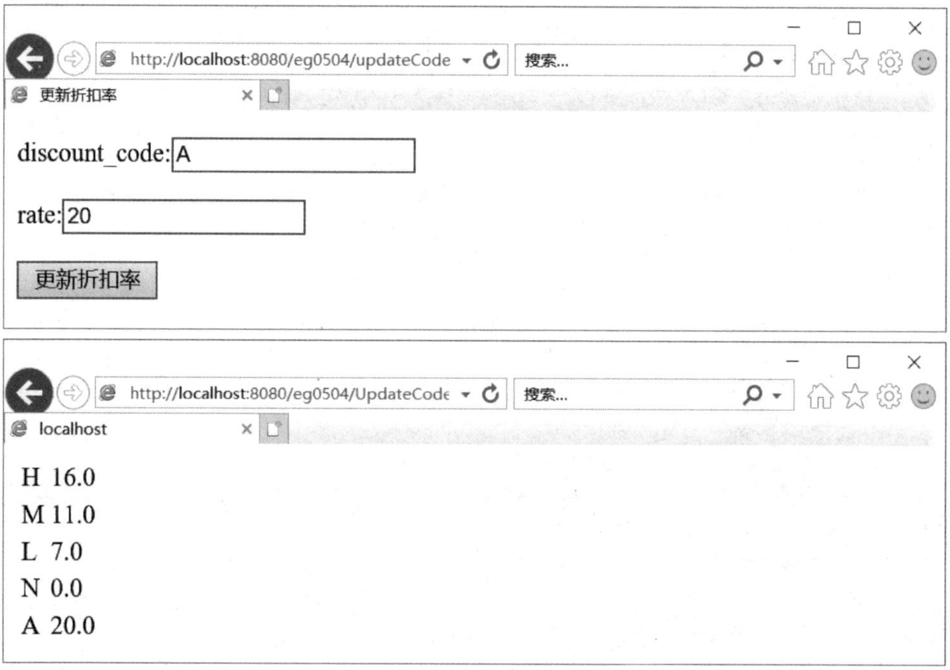

http://localhost:8080/eg0504/updateCode

更新折扣率

discount_code: A

rate: 20

更新折扣率

http://localhost:8080/eg0504/UpdateCode

localhost

H 16.0
M 11.0
L 7.0
N 0.0
A 20.0

图 5.26 折扣码存在时，更新折扣率的运行效果

如果折扣码不存在时，更新折扣率，那么运行效果如图 5.27 所示。

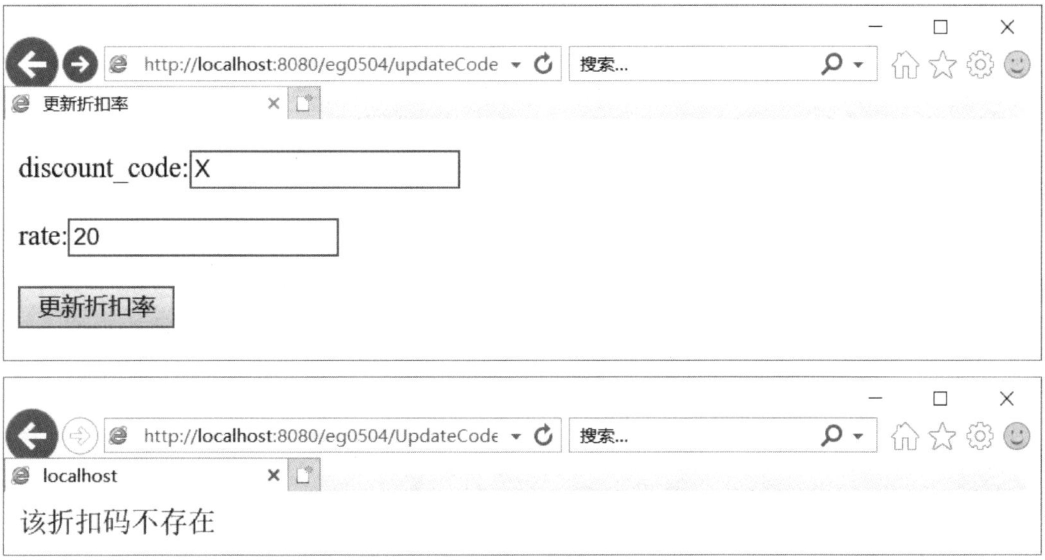

图 5.27 折扣码不存在时，更新折扣率的运行效果

5.9 使用 PreparedStatement 语句对象进行删除操作

首先编写 deleteCode.html 文档的代码，可以复制粘贴 updateCode.html 代码再修改。deleteCode.html

```
<!DOCTYPE html>
<html>
  <head>
    <title>删除折扣码</title>
    <meta charset="UTF-8">
    <meta name="viewport" content="width=device-width, initial-scale=1.0">
  </head>
  <body>
      <form action="DeleteCodeServlet" method="post">
          <p>discount_code:<input type="text" name="discount_code"/></p>
          <p><input type="submit" value="删除折扣码"/></p>
      </form>
  </body>
</html>
```

执行删除操作与执行插入、更新操作类似，可以先复制 UpolateCodeServlet.java 代码再进行改写，大部分自动生成的代码略。DeleteCodeServlet 程序活动图如图 5.28 所示。

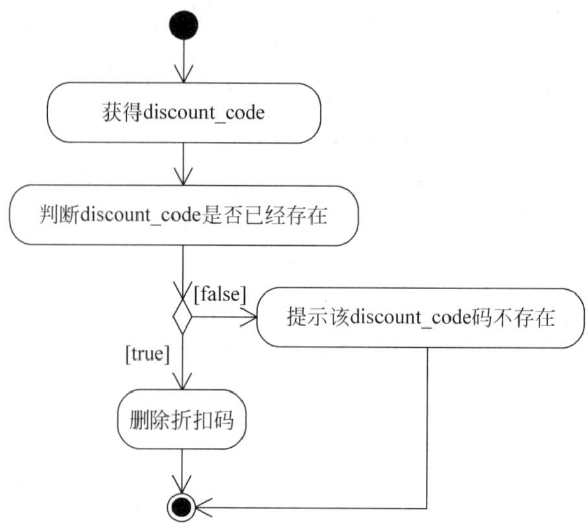

图 5.28 DeleteCodeServlet 活动图

DeleteCodeServlet.java 代码片段

```
package cn.edu.djtu;
//导入包的代码略
@WebServlet(name = "DeleteCodeServlet ", urlPatterns = {"/DeleteCodeServlet "})
public class DeleteCodeServlet extends HttpServlet {
  protected void processRequest(HttpServletRequest request, HttpServletResponse response)
  throws ServletException, IOException {
    response.setContentType("text/html;charset = UTF - 8");
    //获得输入的 discount_code 和 rate
    String discount_code = request.getParameter("discount_code");

    //连接数据库用到的对象
    Connection conn = null;
    PreparedStatement prst = null;
    ResultSet rs = null;

    //连接数据库用到的参数信息
    String url = "jdbc:derby://localhost:1527/sample";
    String driver = "org.apache.derby.jdbc.ClientDriver";
    String user = "app";
    String password = "app";

    //查询数据库的 SQL 语句
    String sql1 = "select discount_code , rate from discount_code where discount_code = ?";
    //删除折扣码的 SQL 语句
    String sql2 = "delete from discount_code where discount_code = ?";

    try (PrintWriter out = response.getWriter()) {
```

```
        Class.forName(driver);
        conn = DriverManager.getConnection(url, user, password);
        prst = conn.prepareStatement(sql1);
        prst.setString(1, discount_code);
        rs = prst.executeQuery();
        if (rs.next()) {
          prst = conn.prepareStatement(sql2);
          prst.setString(1, discount_code);
          prst.executeUpdate();
          request.getRequestDispatcher("ShowAllCodeServlet").forward(request, response);
        } else {
            out.print("该折扣码不存在");
        }
      } catch (ClassNotFoundException | SQLException ex) {
        Logger.getLogger(DeleteCodeServlet.class.getName()).log(Level.SEVERE, null, ex);
      }finally{
        if (rs != null) {
          try {
            rs.close();
          } catch (SQLException ex) {
          Logger.getLogger(DeleteCodeServlet.class.getName()).log(Level.SEVERE, null, ex);
          }
        }

        if (prst != null) {
          try {
            prst.close();
          } catch (SQLException ex) {
          Logger.getLogger(DeleteCodeServlet.class.getName()).log(Level.SEVERE, null, ex);
          }
        }

        if (conn != null) {
        try {
          conn.close();
        } catch (SQLException ex) {
        Logger.getLogger(DeleteCodeServlet.class.getName()).log(Level.SEVERE, null, ex);
        }
      }
    }
      }
      //doGet()、doPost()等方法略
  }
```

如果折扣码存在时，删除折扣码时运行效果如图 5.29 所示。

如果折扣码不存在时，删除折扣码，那么运行效果如图 5.30 所示。

http://localhost:8080/eg0504/deleteCode.

删除折扣码

discount_code: A

删除折扣码

http://localhost:8080/eg0504/DeleteCode

localhost

H 16.0

M 11.0

L 7.0

N 0.0

图 5.29 折扣码存在时,删除折扣码的运行效果

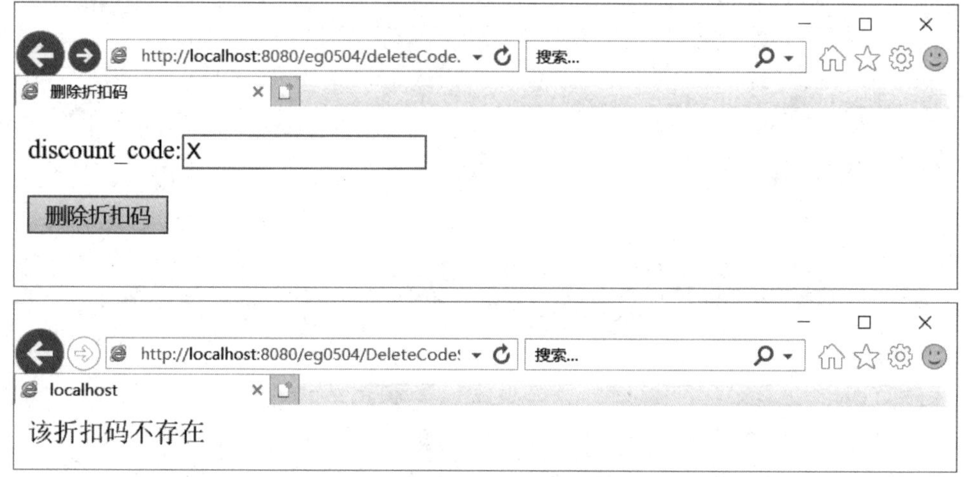

图 5.30 折扣码不存在时,删除折扣码的运行效果

5.10 本章回顾

学习完本章,需要掌握以下内容:

(1) JDBC 的知识。JDBC 定义了一系列的接口,使得程序员可以用统一的模式去操作各种关系数据库。

(2) JDBC 访问数据库的 5 个步骤。

(3) 操纵数据库涉及的对象,这些对象都在 java.sql 包中。包括连接对象、语句对象或预编译语句对象、结果集对象。其中预编译语句对象是语句对象的子接口。

（4）如何载入 JDBC 驱动程序。
（5）如何获得各对象。
（6）预编译语句对象的相关方法。
（7）如何判断结果集是否为空。
（8）如何执行增、删、改操作。

本章小结如图 5.31 所示。

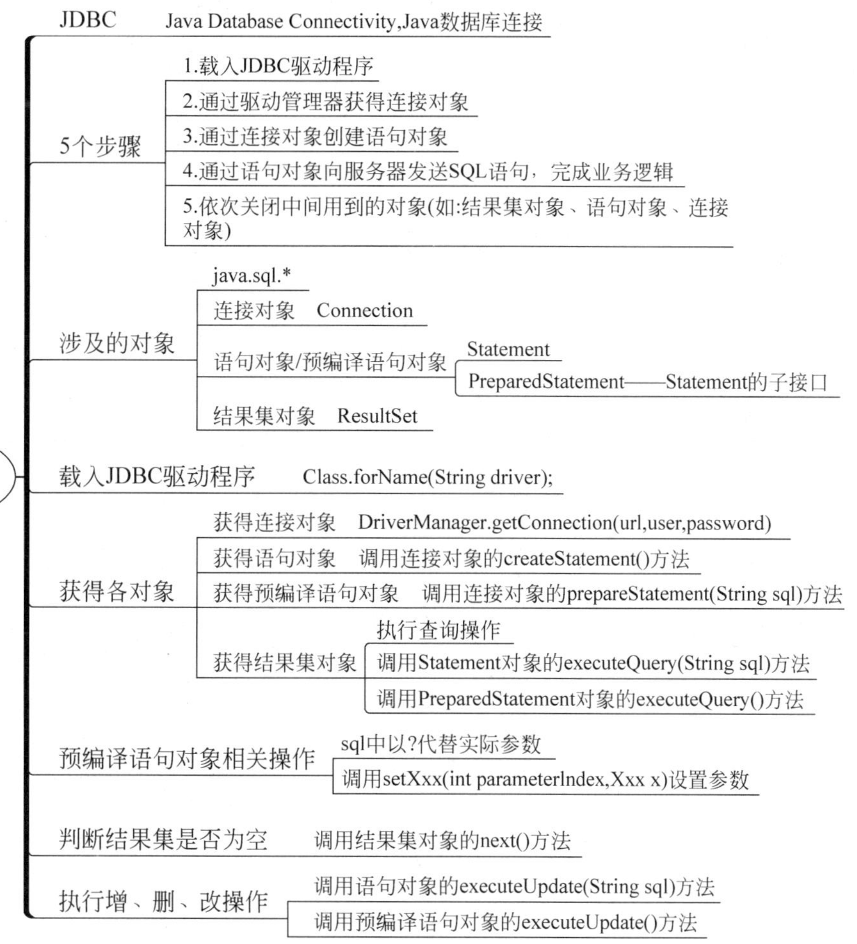

图 5.31 第 5 章内容结构图

5.11 课后习题

1. 名词解释：JDBC。
2. 简述通过 JDBC 访问数据库的 5 个步骤。
3. 如何载入 JDBC 驱动程序？
4. 如何获得连接对象？
5. 如何获得语句对象？

6. 如何获得预编译语句对象？
7. 如何通过预编译语句对象设置 SQL 中的参数？
8. 简述语句对象与预编译语句对象的区别与联系。
9. 如何获得结果集对象？如何判断结果集是否为空？
10. 如何通过语句对象执行增删改操作？
11. 如何通过预编译语句对象执行增删改操作？
12. 改写 eg0501 的例子，使用预编译语句(PreparedStatement)对象完成原来的功能。

第6章 JSP基础

学习目标：

通过本章的学习，你应该：

- 掌握 Servlet 与 JSP 各自的优缺点
- 掌握 Servlet 与 JSP 的关系
- 了解 JSP 页面的组成元素
- 掌握 page 指令的常用属性
- 掌握 JSP 的隐含对象
- 了解如何在 JSP 页面中编写 Java 代码，完成对数据库操作

6.1 JSP 概述

1. Servlet 的优点与缺点

对于 Java 程序员来说，使用 Servlet 很容易上手，编写代码时与编写桌面应用程序差别不大。使用 Servlet 可以处理用户提交的显式或隐式的数据，可以对请求进行转发，对响应进行重定向，对会话进行跟踪，还可以与数据库进行交互。但是，如果细心观察就会发现，Servlet 的这些操作几乎都是与业务逻辑相关的。对于需要呈现给用户的交互界面，Servlet 并不擅长。在第 5 章中，为了输出网页文件，大量使用了 out.print("");语句以使交互界面更友好。例如：

```
out.print("<table>");
  while (rs.next()) {
    out.print("<tr>");
      out.print("<td>" + rs.getInt("customer_id") + "</td>");
      out.print("<td>" + rs.getString("name") + "</td>");
    out.print("</tr>");
  }
out.print("</table>");
```

总结起来，Servlet 的优点主要体现在数据操作等以下几方面：

(1) 读取用户提交的信息。

(2) 生成服务器响应。

(3) 进行会话跟踪。

(4) 对数据库进行操作。

Servlet 的缺点主要体现在数据呈现方面：

(1) 需要大量使用 print 语句生成网页，还要小心处理转义字符。

(2) 专门的网页制作工具不适用于 Java Web 项目。

(3) 要求程序员既要熟悉 Java，又要熟悉网页设计。

那么有没有一种方法，能够既发挥 Servlet 的长处，又同时克服 Servlet 的缺点呢？JSP 技术在一定程度上可以解决这个问题。

2. JSP 的产生及优点

JSP 的产生是市场推动的结果。软件项目的规模越来越大，在需要大量与用户进行交互的项目中，单纯使用 Servlet 的效率比较低下，对开发人员的要求也比较高。于是 JSP 就应运而生。

在 JSP 产生以前，微软公司推出了自己的动态网页技术 ASP，通过在 HTML 的代码中嵌入 VBScript 脚本语言的形式，解决了显示和业务逻辑分离的问题。ASP 技术的市场占有率非常高，开发效率在当时也很高。JSP 提供的技术方案与 ASP 很相似，只是采用的编程语言不是 VBScript 而是 Java。

JSP 很好地解决了 Servlet 所不能解决的问题，它的优点主要体现在这样几个方面：

(1) 不再需要使用 print 语句生成网页，直接用 HTML。

(2) 可以使用专门的网页制作工具来生成网页，极大地提高了开发效率。

(3) 在项目中可以进行分工，将开发人员分成 Java 程序员和网页设计师，让他们在各自擅长的领域内工作。

3. JSP 与 Servlet 的关系

JSP 克服了 Servlet 的缺点，Servlet 不胜任的正是 JSP 所擅长的。从技术演进的角度讲，新技术总会替代旧技术，那么可能有人会问：JSP 是 Servlet 的替代技术吗？

结论是：JSP 并不是 Servlet 的替代技术。虽然不使用任何一个 Servlet，仅用 JSP 也可以完成一个 Web 项目，但是 JSP 并不能完全代替 Servlet。JSP 设计之初并不是为了替代 Servlet，两者侧重点不同，但有着无法割裂的联系。

从本质上讲，每一个 JSP 页面最后都会被 Web 容器自动转换成 Servlet，最终工作的都是 Servlet。但是采用 JSP 技术，可以极大地提高开发效率。在需要向用户输出时，程序员不需要使用 print 语句，也不用关心转义字符等问题。这些工作都可以交给网页设计师去完成。程序员只需要在网页设计师设计好的网页中插入自己需要的动态内容就可以了。

在开发过程中，认为只使用 Servlet 就足够或者只使用 JSP 就足够，这样的想法都是错误的，两者需要相互配合使用。

之前已经学习了 Servlet 的使用，接下来学习 JSP 的一些基础知识，首先看看 JSP 是如何工作的。

6.2 JSP是如何工作的

JSP的设计思想是什么呢？要回答这个问题，还得先从Servlet说起，用Servlet来进行Web开发，从思想上看像是在Java代码中嵌入HTML标记(通过print语句实现)；而JSP则恰恰相反，可以看作在HTML标记中嵌入Java代码(通过脚本标记<% %>来实现)。虽然看起来都是Java代码和HTML混合的形式，但是效果却完全不同。按照JSP的这种方式混合，在一定程度上实现了业务逻辑与数据呈现的分离。

虽然JSP文件的后缀为.jsp，但是JSP页面在修改后第一次访问时，会由Web容器将其转化成一个符合Servlet规范的.java文件，然后再编译成后缀为.class的字节码文件，最后执行字节码文件。

下面通过一个例子来看一下这两者之间的对应关系。

目标：运行一个JSP页面，观察容器生成的与之对应的Servlet。

工程名：eg0601。

用到的文件如表6.1所示。

表6.1 eg0601用到的文件及文件说明

文件名	说 明
index.jsp	为手动创建的JSP文件(注意与index.html对照)

打开NetBeans IDE，新建一个名为eg0601的Java Web项目，再在“Web页”下创建一个名为index.jsp的文档。具体过程略，index.jsp代码如下：

index.jsp

```
<%--
    Document   : index
    Created on : 2018-10-10, 20:59:27
    Author     : Admin
--%>

<%@page contentType="text/html" pageEncoding="UTF-8"%>
<!DOCTYPE html>
<html>
    <head>
      <meta http-equiv="Content-Type" content="text/html; charset=UTF-8">
      <title>JSP Page</title>
    </head>
    <body>
        <h1>Hello World!</h1>
    </body>
</html>
```

运行该文件，则显示“Hello World!”字样，在index.jsp文件上右击，在弹出的菜单上选择“查看 Servlet”命令，过程如图 6.1 所示。

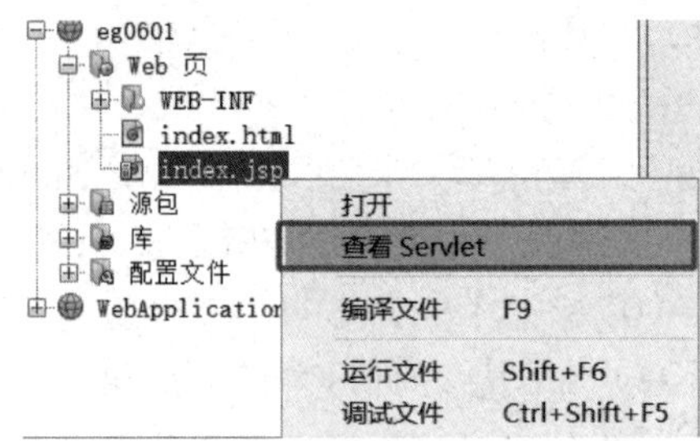

图 6.1 查看 index.jsp 页面生成的 Servlet

在编辑窗口可以见到该文件转换成的 Servlet，此处转成的文件名为 index_jsp.java，如图 6.2 所示。

index_jsp.java × index.jsp ×

C:\Users\Jerry\AppData\Roaming\NetBeans\8.2\config\GF_4.1.1\domain1\generated\jsp\eg0601\org\apache\jsp\index_jsp.java

```
package org.apache.jsp;

import javax.servlet.*;
import javax.servlet.http.*;
import javax.servlet.jsp.*;

public final class index_jsp extends org.apache.jasper.runtime.HttpJspBase
    implements org.apache.jasper.runtime.JspSourceDependent {

  private static final JspFactory _jspxFactory = JspFactory.getDefaultFactory();
```

图 6.2 index.jsp 页面生成的 Servlet——index_jsp.java

转换成的代码如下：

```
package org.apache.jsp;

import javax.servlet.*;
import javax.servlet.http.*;
import javax.servlet.jsp.*;

public final class index_jsp extends org.apache.jasper.runtime.HttpJspBase implements org.
apache.jasper.runtime.JspSourceDependent {
  //代码略
  public void _jspService(HttpServletRequest request, HttpServletResponse response) throws
java.io.IOException, ServletException {
    PageContext pageContext = null;
    HttpSession session = null;
    ServletContext application = null;
    ServletConfig config = null;
    JspWriter out = null;
    Object page = this;
```

```
    JspWriter _jspx_out = null;
    PageContext _jspx_page_context = null;

    try {
      response.setContentType("text/html;charset = UTF - 8");
      response.setHeader("X - Powered - By", "JSP/2.3");
      pageContext = _jspxFactory.getPageContext(this, request, response, null, true, 8192,
true);
      _jspx_page_context = pageContext;
      application = pageContext.getServletContext();
      config = pageContext.getServletConfig();
      session = pageContext.getSession();
      out = pageContext.getOut();
      _jspx_out = out;
      _jspx_resourceInjector = (org.glassfish.jsp.api.ResourceInjector) application.
getAttribute("com.sun.appserv.jsp.resource.injector");
      out.write("\n");
      out.write("\n");
      out.write("\n");
      out.write("<!DOCTYPE html>\n");
      out.write("<html>\n");
      out.write("    <head>\n");
      out.write("<meta http - equiv = \"Content - Type\" content = \"text/html; charset = UTF
- 8\">\n");
      out.write("        <title>JSP Page</title>\n");
      out.write("    </head>\n");
      out.write("    <body>\n");
      out.write("        <h1>Hello World!</h1>\n");
      out.write("    </body>\n");
      out.write("</html>\n");
    } catch (Throwable t) {
       //代码略
    } finally {
       _jspxFactory.releasePageContext(_jspx_page_context);
    }
    }
}
```

从 index_jsp.java 代码中可以看到，index.jsp 中的

```
<%@page contentType = "text/html" pageEncoding = "UTF - 8" %>
```

转换成了

```
response.setContentType("text/html;charset = UTF - 8");
```

而其他的 HTML 标记，转换成了类似

```
out.write("< html >\n");
```

这样的输出语句。

了解了JSP是如何工作的,那么接下来就一起看看JSP页面的组成。

6.3 JSP页面的组成

JSP页面分成两个部分:静态部分和动态部分。

静态部分称之为模板文本(template text),一般由HTML元素构成,这部分JSP容器是不处理的,会原封不动地由服务器端传递到客户端。

动态部分与调用Java代码有关,根据调用方式的不同,主要有指令标记、声明标记、脚本标记、表达式标记、动作标记和注释标记等。在这些标记中,随着项目规模的增大,考虑到维护性的因素,声明标记、脚本标记在页面上已经很少出现了,甚至在有些项目中,明令要求"在JSP页面上严禁直接出现任何Java脚本"。表达式标记虽然因为输出动态内容的需要得以保留,但是在JSP 2.0版本以后,一般也由更为便利的EL(表达式语言)替代(关于EL将在第8章讲解)。所以在学习中,声明标记、脚本标记和表达式标记这三个部分无须深入,只侧重掌握指令标记、动作标记和注释标记即可。

下面将分别学习JSP页面中的各个组成部分。在学习中并不是所有的标记都要掌握,只要会使用常用的标记就可以了,如果有特别的需求,可以查阅相关API。

1. HTML标记(模板文本)

HTML负责生成JSP页面的静态部分(模板文本),这部分代码将原封不动地从服务器端传递到客户端。在客户端浏览器中右击,选择查看源文件,能看到源代码,与服务器端看到的源代码对应部分一模一样。

2. JSP指令标记

JSP指令标记主要用于设定JSP页面的整体配置信息,以便于在转换期间将JSP页面顺利转换为Servlet。指令标记以<%@起始,以%>结束,语法形式如下:

```
<%@  JSP标准指令  属性1  属性2  …  属性n  %>
```

JSP指令标记分为page、include和taglib三种。

1) page指令

page指令可以设置类的导入、内容类型、是否使用会话对象、是否是错误页等。page指令可以放在JSP页面的任何地方,但是通常为了程序的可读性,在编程时都放在顶部位置。

page指令可以定义下面这些属性(大致按照使用的频率列出):import、contentType、pageEncoding、session、isELIgnored(JSP 2.0中支持)、buffer、autoFlush、info、errorPage、isErrorPage、isThreadSafe、language、extends,如表6.2所示。其中前三个(import、contentType、pageEncoding)使用较多,其他的使用较少。

表 6.2 page 指令常用属性

属性	作用
import	import 属性用来指定 JSP 需导入哪些 Java 组件或类。例如，想要在页面中使用 java.util.Date 类，则需要这样设置 page 指令的 import 属性 <%@ page import = "java.util.Date" %> 在转换时，该行语句会转换为 import java.util.Date; 一般来说，一个属性只能指定一个特定的值，但是 import 属性比较特殊，它可以使用逗号分隔的形式，一次引入多个类，例如： <%@ page import = "java.util.Date , cn.edu.djtu.*" %> 在转换时，该行语句会转换为 import java.util.Date; import cn.edu.djtu.*;
contentType	contentType 属性用来设置响应的 MIME 类型。例如，设置成网页形式则按如下方式设置 <%@ page contentType = "text/html" %> 设置成 Excel 则按如下方式设置 <%@ page contentType = "application/vnd.ms-excel %>
pageEncoding	pageEncoding 属性用来设置字符编码，在转换期间，告知容器如何处理 JSP 页面中的字符编码，以及转换后如何设置 charset。在 NetBeans IDE 中默认编码为 UTF-8，可以很好地处理亚洲字符集，如下列代码所示： <%@ page pageEncoding = "UTF-8" %> 设置 MIME 类型和字符编码往往同时进行，在 NetBeans IDE 中默认是这样的： <%@ page contentType = "text/html" pageEncoding = "UTF-8" %> 转换成的 Servlet 中对应的代码是 response.setContentType("text/html;charset = UTF-8");

contentType 属性与 pageEncoding 属性的若干问题

容器会将 JSP 文档转化后的 Servlet 文件(.java 的文件)以 UTF-8 格式存储，并以 UTF-8 的编码编译，而 pageEncoding 主要是告诉容器，这个 JSP 文件的文字编码是什么，以便在转化过程中正确转化为.java。但是 pageEncoding 对响应的 MIME 类型也有影响，如果不设置 contentType 而只设定 pageEncoding，例如，按如下代码形式设置，

```
<%@ page pageEncoding = "UTF-8" %>
```

那么转化后的 Servlet 中对应代码为：

```
response.setContentType("text/html; charset = UTF - 8");
```

如果在 JSP 设置 contentType 时未指定 charset，只设定 contentType 为 text/html 而设置 pageEncoding=GB2312 时，例如：

```
<%@ page contentType = "text/html" pageEncoding = "GB2312" %>
```

则转化后的 Servlet 中对应的代码为：

```
response.setContentType("text/html; charset = GB2312");
```

以上的代码看起来问题还不大，但是如果按如下方式设置，

```
<%@ page contentType = "text/html; charset = GB2312" pageEncoding = "UTF - 8" %>
```

那么转化后的 Servlet 中对应的代码为：

```
response.setContentType("text/html;charset = GB2312");
```

可见 contentType 的设置对响应起了作用。

为了避免困扰，建议在 contentType 中只设置 MIME，在 pageEncoding 中设置字符编码，代码如下：

```
<%@ page contentType = "text/html" pageEncoding = "UTF - 8" %>
```

或者将两者设置成相同的字符编码，代码如下：

```
<%@ page contentType = "text/html; charset = UTF - 8" pageEncoding = "UTF - 8" %>
```

这两种方式在转化后的 Servlet 中，代码一致，均为：

```
response.setContentType("text/html;charset = UTF - 8");
```

浏览器会据此来解释相应的符合 HTML 编码规范的文件。

除此之外，在 JSP 页面上，有时候还有一个地方的编码设置会让大家感到困惑，那就是 meta 标签，例如：

```
<meta http - equiv = "Content - Type" content = "text/html; charset = UTF - 8">
```

这里面也设置了内容类型和字符编码，它什么时候起作用呢？其实，这个设置也是浏览器在对响应进行解释时供参考用的，浏览器在解释服务器发来的响应文件时，优先参考

```
response.setContentType("text/html;charset = UTF - 8");
```

而 meta 标签次之。

page 指令的多个属性是可以写在一条语句里的，例如这样：

```
<%@ page import = "java.util.Date" contentType = "text/html" pageEncoding = "UTF - 8" %>
```

上面的语句中设置了三个属性，还可以继续设置其他属性，但是通常不这样做，因为一次设置多个属性的话，代码的可读性就降低了。在开发过程中，一般只有设置 MIME 类型和字符编码的属性时写在一条语句里，其他属性还是各自设置。

page 指令其他的属性一般不主动设置，了解即可，如表 6.3 所示。

表 6.3 page 指令的其他属性

属性	作　　用
session	指定 JSP 是否包含在 HTTP 会话 session，属性值可设定为 true 或 false，默认值为 true。 `<%@ page session = "true" %>`
isELIgnored	设定 JSP 网页中是否忽略 EL(表达式语言)，默认值是 false，如果设置为 true，那么将不转译表达式语言。该设置会覆盖 web.xml 中的< el-ignored >设置

buffer	设定输出缓冲区的大小，可以设置为 none 或者一个确定的值，单位是 Kb，不小于 8Kb。
	当需要 Servlet 直接输出到输出对象时，赋值为 none，如下所示。
	`<%@ page buffer = "none" %>`
	设置缓冲区为 8Kb 的语句：
	`<%@ page buffer = "8Kb" %>`

autoflush	当输出缓冲区溢出时，是否自动将输出结果传回客户端。默认值是 true，如果设置为 false，则溢出时会发生异常。
	`<%@ page autoFlush = "true" %>`
	通常 buffer 和 autoflush 属性在一起设置，例如：
	`<%@ page buffer = "16Kb" autoflush = "true" %>`

info	用于设定当前 JSP 页面的基本信息，该信息可在转换成的 Servlet 中用 getServletInfo()取得。 `<%@ page info = "This is a test" %>`
errorPage	指定当 JSP 抛出异常时，应该导向哪个“错误页面”(error page)。 `<%@ page errorPage = "error.jsp" %>`
isErrorPage	指定 JSP 是否为“错误页面”。 `<%@ page isErrorPage = "true" %>`

续表

属性	作　　用
isThreadSafe	设置 JSP 页面是否是线程安全的，默认值是 true。假如设置为 false，那么在执行当前 JSP 页面时将只有一个线程。其实通常都认为 JSP 页面是线程安全的，如果设置成 false 会影响整个应用的性能，所以建议保持默认值 true。 `<%@ page isThreadSafe = "true" %>`
language	设置在脚本标记中使用的脚本语言，毫无疑问，这个通常都是设置成 java。 `<%@ page language = "java" %>`
extends	该属性指明了转译生成的 Servlet 必须继承的父类，例如，在 JSP 容器中，JSP 转化成的 Servlet 默认继承自 HttpJspBase，而 HttpJspBase 又继承自 HttpServlet，基本上不会用到这个属性。以下代码表示转译生成的 Servlet 继承自 cn.edu.djtu 包中的 SomeServlet 类(前提是 SomeServlet 必须是一个 Servlet，否则编译失败)。 `<%@ page extends = "cn.edu.djtu.SomeServlet" %>`

2）include 指令

include 指令一般称为“静态包含”，它允许 JSP 页面上在该指令所在的位置插入被包含的内容，两部分内容合并之后再转化成一个 Servlet。语法形如：

```
<%@include file = "somefile" %>
```

虽然 include 指令的功能非常强大，但是带来的维护问题也同样严重。所以，在实际使用中并不推荐使用 include 指令，除非特殊需求，否则，应该尽可能使用“动态包含”来代替。即动作标记中的

```
<% jsp:include page = "somepage"/>
```

3）taglib 指令

JSP 页面中除了可以使用标准动作标记，还支持用户自己扩展标记库，这时候就需要用到 taglib 指令。

```
<%@taglib uri = " " prefix = " " %>
```

其中 uri 是指标记库的 URI，prefix 是指标记的前缀。taglib 指令一般也是放在 JSP 页面的顶部。在第 8 章将会用到它，此处暂不展开。

3. JSP 声明标记、脚本标记、表达式标记

通过前面章节的学习，已经知道 JSP 页面最终要转化成一个 Servlet，而且也知道通过 JSP 指令标记可以在转换期间做一些必要的设置。那么 JSP 声明标记、脚本标记、表达式标记这三者又各自起什么作用呢？(这一部分内容除了表达式标记偶有应用，基本从 JSP 页面销声匿迹了。而表达式标记也有更好的 EL 作为替代品)。

在一个类里面，可以声明属性和方法，并且在某些方法中可以有输出语句。那么 JSP 页面中是如何与一个类里面的这些部分对应的呢？如表 6.4 所示。

表 6.4 标记与 Servlet 中代码的对应关系

标记	语法	对应 Servlet 中的部分
声明标记	<%!　　%>	类的属性或者方法
脚本标记	<%　　%>	_jspService()中的内容
表达式标记	<%=　　%>	输出部分，等价于输出语句

由此可见，这三者所起的作用就在于允许程序员在 JSP 页面中完成以前在 Servlet 中所做的各种操作。

1）JSP 声明标记

在 JSP 声明标记可以声明变量、方法和类，当 JSP 页面运行时，它们将分别转化成对应 Servlet 的属性、方法和内部类。JSP 声明标记可以放在页面的任何位置，均不会影响其效果，因为转化后它们都属于类的成员，但是习惯上一般将其置于页面头部。

例如，如下代码声明了一个 String 类型的变量 name 和一个无返回值的有参方法 setName：

```
<%!
    //声明一个 String 类型的变量 name
    private String name;
    //声明一个无返回值的有参方法 setName
    public void setName(String name) {
        this.name = name;
    }
%>
```

2）JSP 脚本标记

JSP 脚本标记有时候被称为 Java 脚本片段。在脚本标记中写的代码转换后将变成_jspService()中的代码。JSP 脚本标记在 JSP 页面中可以多次使用，将按照出现顺序依次转入_jspService()中。

例如，以下代码运行将输出 5 个 Hello，尽管看起来 for 循环体被中间网页上的文字隔断成两个片段，但是转化后 3 个部分都变成_jspService()中的代码了。

```
<%
    for (int i = 0; i < 5; i++) {
%>
    Hello
<%
    }
%>
```

转换后的对应部分大致是这样的：

```
for (int i = 0; i < 5; i++) {
    out.write("\n");
    out.write("            Hello!\n");
    out.write("        ");
}
```

3）JSP 表达式标记

JSP 表达式标记的作用是输出表达式的运行结果，只要该表达式有值。

例如：

```
<%= "hello world!" %>
```

转换后的对应部分大致是这样的：

```
out.print("hello world!");
```

下面通过一个例子将声明标记、脚本标记、表达式标记三者结合起来做一个简单的例子，例子不足取，旨在理解三者在 JSP 页面中的作用。

目标：加深对声明标记、脚本标记、表达式标记的理解。

工程名：eg0602。

用到的文件如表 6.5 所示。

表 6.5　eg0602 用到的文件及文件说明

文件名	说　明
index.jsp	为创建工程时自动创建的文件。创建后在其中添加所需代码

打开 NetBeans IDE，新建一个名为 eg0602 的 Java Web 项目，具体过程略；index.jsp 代码如下，创建工程时自动生成的代码，创建后在其中添加了所需代码：

index.jsp

```
<%@page contentType = "text/html" pageEncoding = "UTF-8" %>
<!DOCTYPE html>
<html>
    <head>
<meta http-equiv = "Content-Type" content = "text/html; charset = UTF-8">
        <title>声明标记、脚本标记、表达式标记的使用</title>
    </head>
    <body>
        <%!
            private String name;
            public void setName(String name) {
```

```
                this.name = name;
            }
        %>
        <%
            if (this.name == null || "".equals(this.name)) {
                this.setName("青椒麦客");
            }
        %>
        <% = name %>
    </body>
</html>
```

运行之后在网页上将显示“青椒麦客”。此时在index.jsp文件上右击，在弹出的菜单上选择“查看Servlet”命令(具体过程参见eg0601)。在编辑窗口可以见到该文件转换成的Servlet，此处转成的文件名为index_jsp.java，节选部分代码如下：其中与本节内容无关的略去了，对应转化部分以波浪线形式提示。

index_jsp.java代码片段

```
//代码略
public final class index_jsp extends org.apache.jasper.runtime.HttpJspBase implements org.
apache.jasper.runtime.JspSourceDependent {

    private String name;

    public void setName(String name) {
        this.name = name;
    }
      //代码略
  public void _jspService(HttpServletRequest request, HttpServletResponse response) throws
java.io.IOException, ServletException {
    try {
        //代码略
      if (this.name = = null || "".equals(this.name)) {
          this.setName("青椒麦客");
      }
        //代码略
      out.print(name);
  }
    //代码略
  }
}
```

从例子可以看到，如果想要完成的操作等价于声明类的属性和方法，那么在JSP页面上就用声明标记<%! %>实现；如果想要完成的操作等价于在_jspService()方法中实现一些业务逻辑，那么就用脚本标记<% %>实现；如果想要完成的操作等价于输出某些信息，那么可以使用表达式标记<%= %>。

但是这样的应用模式目前已经不适用了。因为这样会导致页面上的脚本标记过多，业务逻辑和信息的呈现混杂在一起，当业务逻辑复杂时，项目很难维护。所以催生出了模式一(JSP+JavaBean)这种开发模式，目的是用业务逻辑将信息的呈现分离开，尽可能减少JSP页面上的Java脚本，同时增强代码的复用性。

模式一中提到的JavaBean是JSP动作标记中的一个，因为相对重要，所以放在本书的第7章讲解。

4. JSP动作标记

JSP动作标记设计的初衷是减少页面Java脚本的使用，将原来页面上的一系列操作封装成标记，从而在需要类似操作时，可以只写标记，这样增强了代码的可维护性和重用性。

JSP动作标记分成标准动作和扩展动作，标准动作可以在JSP页面上直接使用。标准动作标记一般形如：

```
<jsp:xxx 属性 1 = "   " 属性 2 = "   ">     </jsp:xxx>
```

扩展动作指的是用户自定义的标记库，在使用时需要先设置taglib指令，然后才能在JSP页面上使用。

JSP 2.0规范中定义了5大类共计20种标准动作，但是从使用效果上看，真正使用它们进行企业级开发的并不多，因此本书中不一一列出，仅介绍其中的几个常用标记，例如：<jsp:include />、<jsp:forward />、<jsp:param />，在本书的第7章介绍JavaBean相关标记<jsp:useBean />、<jsp:setProperty />、<jsp:getProperty />。

1) <jsp:param/>

<jsp:param />一般嵌套在<jsp:include />、<jsp:forward />、<jsp:params />代码块中，作用是在请求过程中传递参数，例如：如下代码传递了一个名为msg的参数，参数值为hello：

```
<jsp:param name = "msg" value = "hello" />
```

2) <jsp:include />

在介绍include指令标记时，已经知道了include指令是“静态包含”，它是先合并被包含的内容，然后转化成一个Servlet。而动作标记中的<jsp:include />是“动态包含”，语法形如：

```
<jsp:include page = "somepage" />
```

两者的区别在于：使用动态包含时，当前JSP页面和被包含的JSP页面各自转化成一个Servlet，在运行期间，将运行结果包含进来。更为灵活的是，还可以在包含期间向被包含页面传递参数。例如，假如a.jsp中有如下代码片段，则在运行期间a.jsp向b.jsp传递了一个参数，参数名为username，参数值为admin。代码如下：

```
<jsp:include page="b.jsp">
    <jsp:param name="username" value="admin" />
</jsp:include>
```

在实际运行过程中，a.jsp 和 b.jsp 各自转化成一个 Servlet。a.jsp 转化成的 Servlet 中会取得 org.apache.jasper.runtime.JspRuntimeLibrary 对象，并执行该对象的 include()方法，将请求转发给 b.jsp 转成的 Servlet，待其执行完毕后，再将控制权转回 a.jsp。这样做的好处是可以组合两个页面的输出结果。

3) <jsp:forward/>

如果只想把请求转发给另一个 JSP 页面，就应该使用<jsp:forward/>标记，该标记用法与<jsp:include />类似，例如，把上一例代码中的 include 都换成 forward。

```
<jsp:forward page="b.jsp">
    <jsp:param name="username" value="admin" />
</jsp:forward>
```

那么在实际运行过程中，a.jsp 和 b.jsp 依然各自转化成一个 Servlet 。a.jsp 转化成的 Servlet 中会取得 PageContext 对象，执行该对象的 forward()方法，将请求转发给 b.jsp 转成的 Servlet，待其执行完毕后，控制权不再转回 a.jsp。这样做只能得到 b.jsp 的输出，a.jsp 中不能有输出，即使有，页面上也显示不出来。

结论：<jsp:include />、<jsp:forward />两个动作标记，运行效果等价于在 Servlet 中调用请求转发器 RequestDispatcher 对象的 include()方法和 forward()方法。在请求转发期间，可以同时使用<jsp:param>传递参数。考虑到这属于业务逻辑跳转的部分，所以还是放在 Servlet 中采用请求转发器 RequestDispatcher 来进行转发更为合适。

5. JSP 注释标记

在代码中添加注释，可以增强代码的可读性，增强程序的可维护性，所以应该适当地添加注释标记。因为 JSP 的组成中有 HTML 标记和 Java 脚本标记，所以这两种语言中的注释标记在 JSP 页面中依然可以使用，例如：

HTML 中的注释(这部分代码在客户端网页上右击，查看源文件是可见的)

```
<!-- HTML 中的注释 -->
```

Java 中的注释，分为单行注释和多行注释两种。这部分代码在客户端是不可见的。

```
<%
    //Java 中的单行注释
    /* Java 中的多行注释 */
%>
```

此外 JSP 页面有自己的注释，这部分代码在客户端也是不可见的。

```
<%-- JSP的注释 --%>
```

容器在将JSP转换成Servlet时,会忽略<%--和--%>之间的内容。

6.4 JSP的隐含对象

在Servlet中进行编程时,NetBeans IDE默认生成的代码中有如下代码,这段代码提供了一个PrintWriter类型的对象,以便于向客户端输出信息:

```
try (PrintWriter out = response.getWriter())
```

曾经在会话管理时写过这样的代码,以便访问一个会话对象:

```
HttpSession mySession = request.getSession();
```

为了更高效地编程,我们频繁使用一些已经准备好的对象,调用相关的方法来完成自己的业务逻辑。

JSP技术设计时也考虑了这些问题,所以已经在JSP页面中准备好了一些对象,这些对象被称作JSP隐含对象。JSP隐含对象由Servlet容器初始化,可以在页面上直接使用。下面看一个JSP页面转化成Servlet之后的代码片段:

```
//代码略
public final class index_jsp extends org.apache.jasper.runtime.HttpJspBase implements org.
apache.jasper.runtime.JspSourceDependent {
  public void _jspService(HttpServletRequest request, HttpServletResponse response)
  throws java.io.IOException, ServletException {
    //代码略
    PageContext pageContext = null;
    HttpSession session = null;
    ServletContext application = null;
    ServletConfig config = null;
    JspWriter out = null;
    Object page = this;
    JspWriter _jspx_out = null;
    PageContext _jspx_page_context = null;
    //代码略
  }
}
```

从代码片段可以看到,隐含对象都是在_jspService()方法内部定义的,所以使用时可以在脚本标记<% %>和表达式标记<%= %>中直接使用。

JSP隐含对象在特定的时期得到了广泛的应用,但是随着设计模式的应用,脚本标记和表达式标记在页面使用的越来越少,这些隐含对象出场的机会也就随之减少了。因此大致了解一下这些隐含对象即可。

JSP 提供了 9 个隐含对象，下表中列出了转译后各自对应的类型及存取范围，并对它们所起的作用做了简单介绍，如表 6.6 所示。

表 6.6 JSP 页面的 9 个隐含对象

隐含对象名	转译后对应类型	存取范围	作　用
request	HttpServletRequest	request	处理请求
response	HttpServletResponse	page	处理响应
session	HttpSession	session	处理会话
page	this	page	相当于 Java 中的 this
application	ServletContext	application	处理上下文
exception	Throwable	page	只出现于 isErrorPage 设定为 true 的 JSP 页面
config	ServletConfig	page	处理转译后的 Servlet 配置信息
pageContext	PageContext	page	可利用其对页面信息进行封装
out	JspWriter	page	处理输出

在实际使用中，并不是所有对象的使用频率都那么高，如果一定要在页面使用这些对象的话，那么 request、response、session 算是常用的，其他对象使用的并不多，即使是看起来应该频繁使用的 out 对象也是如此，输出的操作有一个很好的替代品——输出表达式<%=　%>。

此外，这里的 out 对象与 Servlet 中的那个 out 对象也不完全相同，Servlet 中使用的 out 变量类型是 PrintWriter，而这里的 out 对象类型是 JspWriter。

6.5 使用纯 JSP 进行数据库操作

通过以上的讲解，相信大家对 JSP 已经有了一个初步的了解，下面感受一下使用纯 JSP（只有 JSP 页面，没有 Servlet 参与）进行开发是什么样子的。例子的业务逻辑与第 5 章的 eg0502 一样，通过一个 HTML 页面提交查询条件，查询并输出满足条件的数据。只不过这一次处理请求的组件不是 Servlet 而是一个 JSP 页面，依然使用 PreparedStatement 来完成。

目标：学习在 JSP 页面中通过 JDBC 对数据库进行有条件查询操作（使用 PreparedStatement 的方式）。

工程名：eg0603。

已知条件如表 6.7 所示。

表 6.7 eg0603 的已知条件

数据库	样例数据库 sample	
表名	customer	（表结构略）
欲查询的字段名	customer_id	数据类型为 INTEGER
	name	数据类型为 VARCHAR(30)
查询条件字段名	discount_code	数据类型为 CHAR(1)

用到的文件如表 6.8 所示。

表 6.8　eg0603 用到的文件及文件说明

文　件　名	说　　明
index. html	创建项目时自动生成,页面改写一下。页面有一个表单,表单内有一个文本字段和一个提交按钮,用来提交用户的 discount_code。表单提交给 querybycode. jsp 处理
querybycode. jsp	手动创建的 JSP 页面,用来取得用户通过表单提交的信息,并按照该信息进行查询、输出查询结果

编程思路:

(1) index. html 页面提交信息给 querybycode. jsp。

(2) querybycode. jsp 连接数据库、查询满足条件的用户是否存在。

(3) 如果用户存在,则格式化输出用户的 customer_id 和 name。

(4) 如果用户不存在则不显示信息。

打开 NetBeans IDE,在 eg0603 的 Java Web 项目中新建名为 querybycode. jsp 的页面,具体过程略。编写各部分代码如下:

index. html

```
<! DOCTYPE html >
< html >
  < head >
    < title >提交 discount_code </title >
    < meta charset = "UTF - 8">
    < meta name = "viewport" content = "width = device - width, initial - scale = 1.0">
  </head >
  < body >
      < form action = "querybycode. jsp" method = "post">
        discount_code:< input type = "text" name = "discount_code" />
        < input type = "submit" value = "提交" />
      </form >
  </body >
</html >
```

querybycode. jsp

```
<% --
    Document     : querybycode
    Created on   : 2018 - 10 - 11, 16:05:49
    Author       : Admin
 -- %>

<% @page import = "java. sql. DriverManager" %>
<% @page import = "java. sql. ResultSet" %>
```

```
<%@page import="java.sql.PreparedStatement"%>
<%@page import="java.sql.Connection"%>
<%@page contentType="text/html" pageEncoding="UTF-8"%>
<!DOCTYPE html>
<html>
    <head>
        <meta http-equiv="Content-Type" content="text/html; charset=UTF-8">
        <title>查询 customer 信息</title>
    </head>
    <body>
    <%
        //获得用户提交的 code
        String discount_code = request.getParameter("discount_code");

        //连接数据库用到的对象
        Connection conn = null;
        PreparedStatement prst = null;
        ResultSet rs = null;

        //连接数据库用到的参数信息
        String url = "jdbc:derby://localhost:1527/sample";
        String driver = "org.apache.derby.jdbc.ClientDriver";
        String user = "app";
        String password = "app";

        //查询数据库的 SQL 语句
        String sql = "select customer_id , name from customer where discount_code = ?";

        Class.forName(driver);
        conn = DriverManager.getConnection(url, user, password);
        prst = conn.prepareStatement(sql);
        prst.setString(1, discount_code);
        rs = prst.executeQuery();
    %>
        <table>
    <%
        while (rs.next()) {
    %>
            <tr>
                <td><%= rs.getString(1) %></td>
                <td><%= rs.getString(2) %></td>
            </tr>
    <%
        }
        if (rs != null) {
```

```
            rs.close();
        }
        if (prst != null) {
            prst.close();
        }
        if (conn != null) {
            conn.close();
        }
    %>
        </table>
    </body>
</html>
```

运行结果略。

从代码中可以看出,与用 Servlet 来实现相比较,逻辑实现部分的代码没什么差别,主要是在格式化输出时变化较大,原来用 print 语句输出的 HTML 标记,现在可以在页面上直接写;原来输出变量的语句可以用表达式标记<%=　%>代替,编程比较灵活。这种灵活性对于小型项目来说是比较适合的,尤其是项目将来不需要扩展或者过多维护时。但是,一旦项目规模变大,这种混合代码的编程风格就不适用了,维护起来比较麻烦,查错也比较困难。因此,这种模式在大型的项目中就基本上没有用武之地,渐渐绝迹了。Java Web 开发中的模式一(JSP+JavaBean)和模式二(JSP+JavaBean+Servlet)也就应运而生了。当然其中模式一的方式现在已经没有什么应用了,本书后续章节主要从模式二开始介绍。

此外还有一点不同就是在 JSP 页面编写 Class.forName()这样的代码时,NetBeans IDE 是不提示抛出异常的,如果想要处理异常需要自己主动加上 try-catch 块,本样例未做处理,以作对比。

6.6 本章回顾

本章一开始对 Servlet 和 JSP 做了比较,讲述了两种技术各自的优缺点,阐述了两者之间的关系;接着讲述了 JSP 的工作原理;介绍了 JSP 页面的组成及隐含对象,最后用 JSP 技术完成了一个数据库操作的例子。就本章的内容而言,真正需要大家动手去掌握的,落实到程序上的并不多,毕竟当前 Web 应用程序的开发模式与 JSP 刚开始使用时有了很大的变化,从前那种在页面上写大量 Java 脚本的日子一去不复返了。

但是学习本章依然是有意义的,通过本章的学习,能够知道 JSP 的设计初衷;通过对比,可以更好地理解 Servlet 和 JSP 两者之间的关系,知道两者各自擅长的是什么。这对于学习模式一与模式二无疑是有好处的。本章小结如图 6.3 所示。

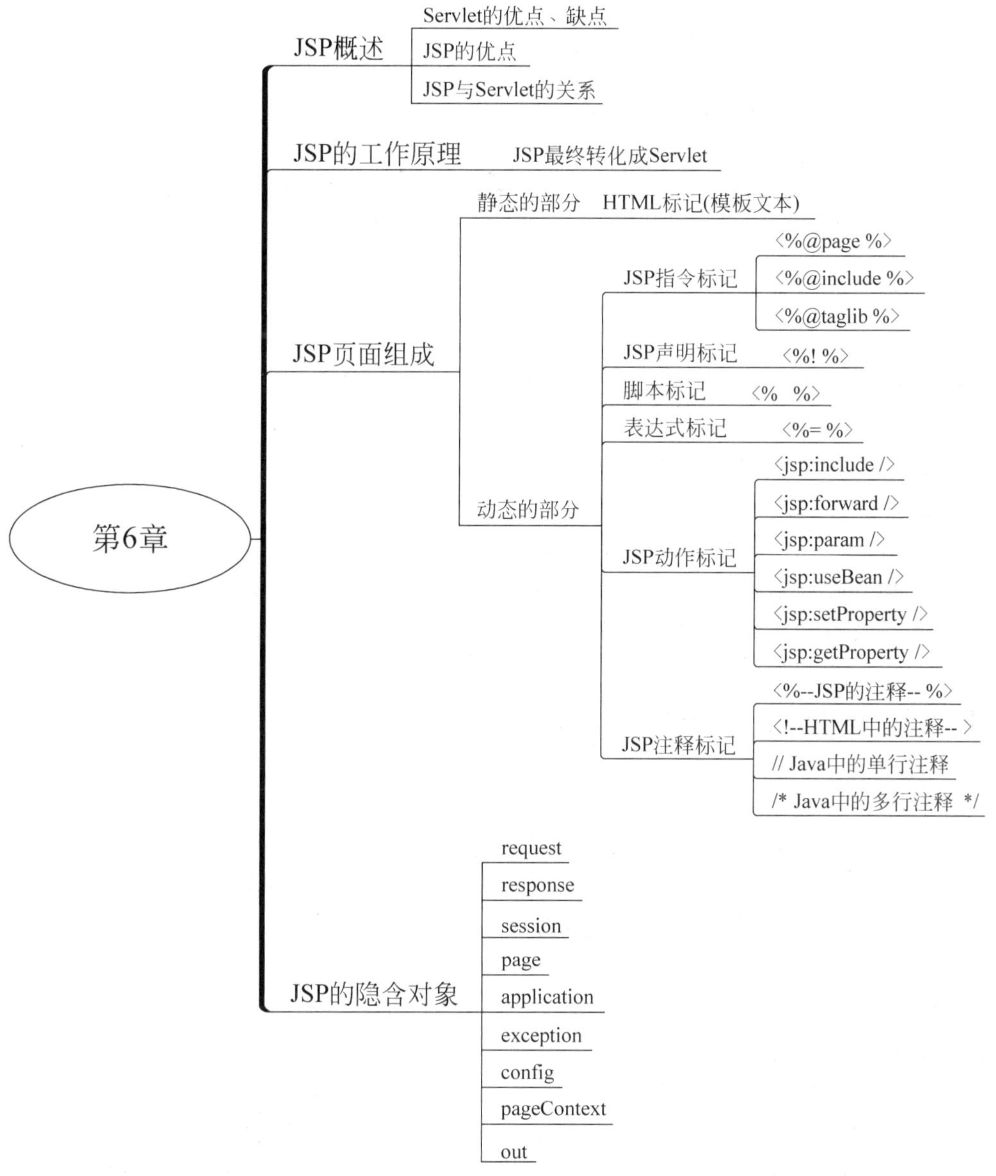

图 6.3　第 6 章内容结构图

6.7　课后习题

1. Servlet 的优点与缺点有哪些？
2. 简述 Servlet 与 JSP 的区别与联系。
3. 简述 JSP 的工作原理。
4. 简述 JSP 页面的组成有哪些？
5. JSP 页面上支持的注释有哪几类？
6. JSP 的常用隐含对象有哪些？分别是什么类型？（至少说出 5 个）。

第7章 JSP与JavaBean

学习目标：

通过本章的学习，你应该：

- 理解什么是 JavaBean，JavaBean 的要求
- 理解在 JSP 页面上如何使用 JavaBean
- 掌握 JSP 中 JavaBean 的存取范围
- 了解<jsp:useBean />、<jsp:setProperty />、<jsp:getProperty />标准动作的用法
- 掌握 Servlet＋JSP＋JavaBean 三者相结合的开发模式

7.1 JavaBean 的定义及语法

1. JavaBean 的定义

JavaBean 这个概念是一个不大容易厘清的概念，因为在不同时期、不同场合它指代的对象不同。本书中接受的 JavaBean 定义如下：JavaBean 是 Java 开发语言中一个可以重复使用的软件组件。

JavaBean 并不是专门为 Web 应用程序设计的。JavaBean 分为两类：可视化的 JavaBean 和非可视化的 JavaBean。

(1) Java 技术体系中的 AWT 或者 Swing 组件，其中的窗体、菜单、按钮、文本框之类的组件都可以看作是 JavaBean。这一类的 JavaBean 侧重交互界面，被称作可视化的 JavaBean。

(2) 在 Web 应用程序中使用的 JavaBean，主要用来封装数据或者业务逻辑。这一类 JavaBean 是不可见的，所以被称作非可视化的 JavaBean。(模式一中使用 JavaBean 时也可以封装一部分业务逻辑。而在模式二中主要用于封装数据)。

2. JavaBean 的语法

JavaBean 本质上就是符合 Java 语法的类，只不过在此基础上，对它还有更为细致的要求。在 Web 应用程序中使用的 JavaBean，一般要满足如下要求：

(1) 必须有一个访问权限为 public 的无参数的构造方法。

(2) 属性的访问权限为 private。

(3) 属性类型必须是 String 或者基本数据类型(也可以是基本数据类型的包装类)。

(4) 属性名和类型是由获取方法(getter)和设置方法(setter)推导出的。

(5) 必须按命名约定来命名访问权限为 public 的获取方法(getter)和设置方法(setter)。

(6) 设置方法(setter)的参数类型和获取方法(getter)的返回类型必须一样。

7.2　编写一个 JavaBean

下面通过一个例子来学习一下 JavaBean 的编写。第 5 章中曾经操纵过 sample 数据库的 customer 表。当时为了简化练习只选取了表中的两个字段 customer_id(数据类型为 INTEGER)、name(数据类型为 VARCHAR(30))。选中 customer 表右击,选择"执行命令",在右侧窗口处输入"SELECT customer_id, name FROM customer"后按 Ctrl+Shift+E 组合快捷键可以执行该 SQL 语句,观察该语句执行结果,如图 7.1 所示。

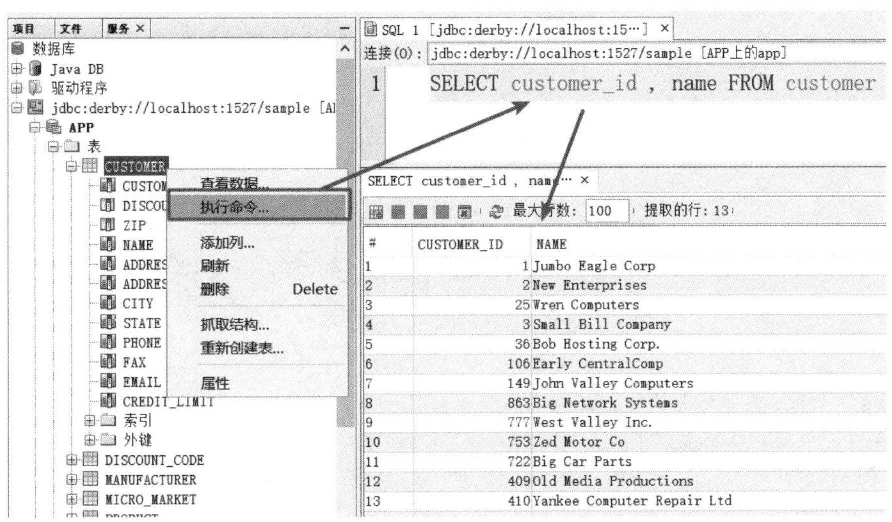

图 7.1　执行查询的 SQL 语句

为了便于描述,现假设 customer 表仅有这两个字段(有多个字段其实也一样)。

按照关系数据库的术语,可以有如下描述:有一张 customer 表,表中有 2 个字段,字段 customer_id 的数据类型为 INTEGER,字段 name 的数据类型为 VARCHAR(30)。表中有 13 条记录,每一条记录对应一个实体,每一个实体都是一个 customer。

那么在 Java 代码中如何描述这件事呢?可以这样去描述:有一个类 Customer,它对应 customer 表。Customer 类有两个属性,一个属性是 customer_id,数据类型为 INTEGER;另一个属性是 name,数据类型为 String(Java 中与 VARCHAR 数据类型对应的是数据类型是 String)。表中的 13 条记录对应着 Customer 类的 13 个实例(或者说对象)。当然,为了设置各个属性并获得各个属性值,还需要编写对应的 getter 方法和 setter 方法。Customer 类的类图如图 7.2 所示。

Customer
–customer_id –name
+getCustomer_id() +setCustomer_id(customer_id) : void +getName() +setName(name) : void

图 7.2　Customer 类的类图

目标:学习编写一个 JavaBean。

工程名：eg0701。

打开 NetBeans IDE,新建一个名为 eg0701 的 Java Web 项目。新建一个名为 Customer 的类,设置类名为 Customer,包名为 cn. edu. djtu. bean。当类越来越多时,一般会按功能不同,将类放到不同的包中,bean 包中放的都是 JavaBean,也有些人会将包命名为 vo(值对象)或者 dto(数据传输对象),如图 7.3 所示。

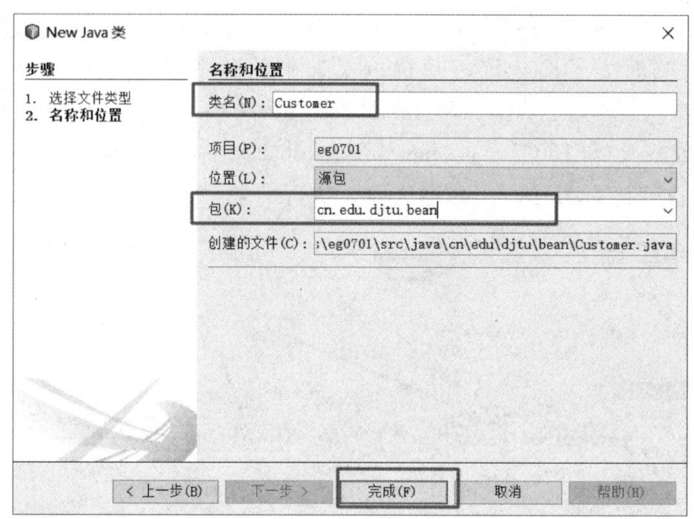

图 7.3 创建 Customer 类

在 Customer 类代码中输入

```
int customer_id;
String name;
```

然后在代码空白处右击,选择"重构"→"封装字段"命令,如图 7.4 所示。

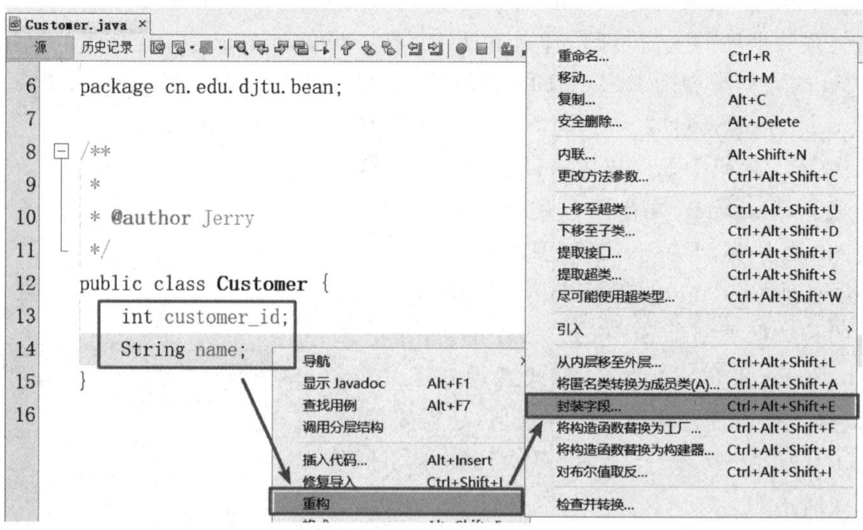

图 7.4 重构-封装字段操作

在弹出的窗口中先后单击“全选”按钮和“重构”按钮，如图7.5所示。

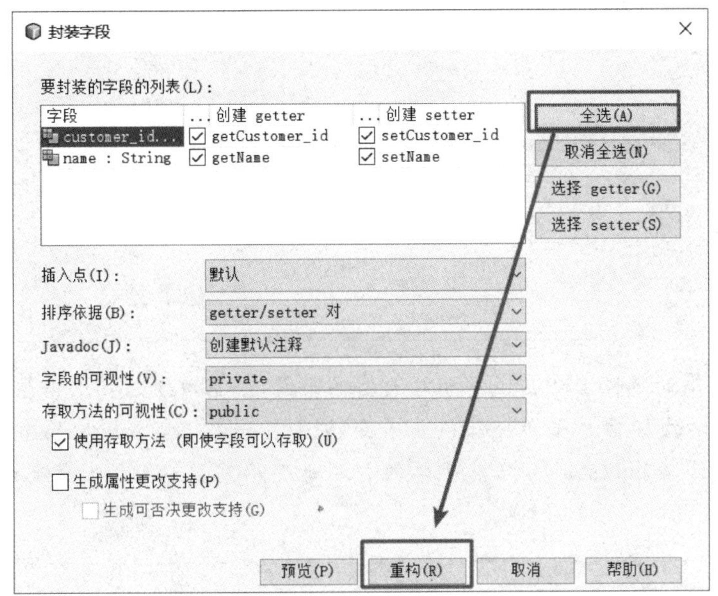

图7.5 全选-重构操作

操作完成后会生成如下代码，此时getter/setter代码都已经自动完成了。

Customer.java代码如下：

```
package cn.edu.djtu.bean;

public class Customer {
    private int customer_id;
    private String name;

    /**
     * @return the customer_id
     */
    public int getCustomer_id() {
        return customer_id;
    }

    /**
     * @param customer_id the customer_id to set
     */
    public void setCustomer_id(int customer_id) {
        this.customer_id = customer_id;
    }

    /**
     * @return the name
     */
```

```
    public String getName() {
        return name;
    }

    /**
     * @param name the name to set
     */
    public void setName(String name) {
        this.name = name;
    }
}
```

注意：如果类的属性名声明成首字母大写的形式，如下所示。那么严格地说，这个类并不符合 JavaBean 的标准。虽然 getter/setter 方法没有区别。（JavaBean 中的属性是由 getter/setter 方法将 get/set 去掉，然后单词首字母变小写得到，所以属性的首字母必然是小写的）。

```
private int Customer_id;
private String Name;
```

7.3 （Servlet+JSP+JavaBean）结合使用案例 1

到目前为，已经学习了 Servlet、JSP 和 JavaBean 三种技术。Servlet 适合控制业务流程，JSP 适合数据显示，JavaBean 适合完成数据封装。下面通过一个案例将三者结合起来。

目标：学习将 Servlet+JSP+JavaBean 结合使用完成数据库查询。Servlet 负责业务流程控制，JSP 负责数据显示，JavaBean 负责数据封装。

工程名：eg0702。

用到的文件如表 7.1 所示。

表 7.1　eg0702 用到的文件及文件说明

文件名	说明
index.html	创建工程时自动创建
ConditionalQueryServlet.java	手动创建的 Servlet 文件，用于完成数据库查询及请求转发
Customer.java	JavaBean 用于完成数据封装
showCustomer.jsp	手动创建的 JSP 文件，用于满足条件的 customer 信息输出

编程思路：

（1）用户通过 index.html 提交一个 customer_id。

（2）ConditionalQueryServlet 根据提交的 customer_id 查询数据库，如果该 customer_id 对应的客户存在，则将其封装成一个 Customer 对象并存入 request 域，将请求转发至 showCustomer.jsp。

(3) 在 showCustomer.jsp 取出该 Customer 对象并输出该对象信息。

(4) 业务逻辑分析如图 7.6 所示。

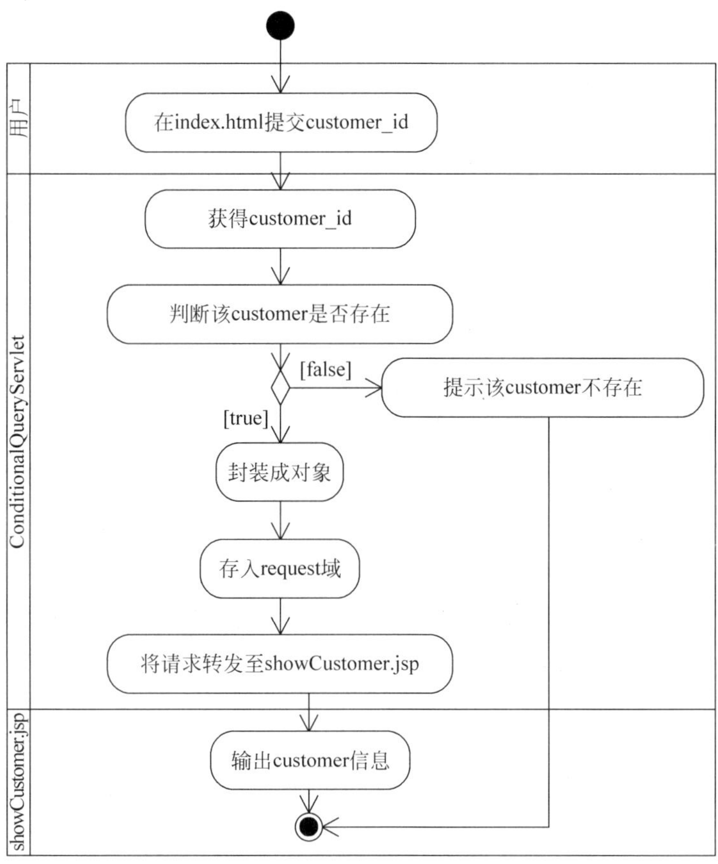

图 7.6　eg0702 业务逻辑分析

为了简化业务逻辑，假设用户一定会录入一个整型的 customer_id，而没有编写相应的前端 JS 校验代码。

打开 NetBeans IDE，新建一个名为 eg0702 的 Java Web 项目。分别新建文件名为 showCustomer.jsp 的 JSP 文件和文件名为 ConditionalQueryServlet.java 的 Servlet 文件，文件名为 Customer.java 的文件，其中 Customer.java 文件代码与 eg0701 中的 Customer.java 完全一样。

Customer.java 代码

```
package cn.edu.djtu.bean;

public class Customer {
    private int customer_id;
    private String name;

    /**
```

```
         * @return the customer_id
        */
       public int getCustomer_id() {
           return customer_id;
       }

       /**
         * @param customer_id the customer_id to set
        */
       public void setCustomer_id(int customer_id) {
           this.customer_id = customer_id;
       }

       /**
         * @return the name
        */
       public String getName() {
           return name;
       }

       /**
         * @param name the name to set
        */
       public void setName(String name) {
           this.name = name;
       }
}
```

index.html 代码也比较简单。

index.html 代码

```
<!DOCTYPE html>
<html>
    <head>
        <title>提交 customer_id</title>
        <meta charset="UTF-8">
        <meta name="viewport" content="width=device-width, initial-scale=1.0">
    </head>
    <body>
        <form action="ConditionalQueryServlet" method="post">
            <p>customer_id:<input type="text" name="customer_id" /></p>
            <p><input type="submit" value="查询" /></p>
        </form>
    </body>
</html>
```

ConditionalQueryServlet.java 代码与之前的查询类似，略有差异的地方见代码中波浪线处。

ConditionalQueryServlet.java 代码片段

```
package cn.edu.djtu;
//代码略
@WebServlet(name = "ConditionalQueryServlet", urlPatterns = {"/ConditionalQueryServlet"})
public class ConditionalQueryServlet extends HttpServlet {
    protected void processRequest(HttpServletRequest request, HttpServletResponse response)
throws ServletException, IOException {
    response.setContentType("text/html;charset = UTF - 8");
    //获得用户提交的 customer_id
    String customer_id = request.getParameter("customer_id");

    //连接数据库用到的对象
    Connection conn = null;
    PreparedStatement prst = null;
    ResultSet rs = null;

    //连接数据库用到的参数信息
    String url = "jdbc:derby://localhost:1527/sample";
    String driver = "org.apache.derby.jdbc.ClientDriver";
    String user = "app";
    String password = "app";

    //查询数据库的 SQL 语句
    String sql = "select customer_id , name from customer where customer_id = ?";

    try (PrintWriter out = response.getWriter()) {
        Class.forName(driver);
        conn = DriverManager.getConnection(url, user, password);
        prst = conn.prepareStatement(sql);
        prst.setInt(1, Integer.parseInt(customer_id));
        rs = prst.executeQuery();
        if (rs.next()) {
            Customer c = new Customer();
            c.setCustomer_id(rs.getInt(1));
            c.setName(rs.getString(2));
            //以 key 名 c,存储了引用名为 c 的 Customer 对象
            request.setAttribute("c", c);
    request.getRequestDispatcher("showCustomer.jsp").forward(request, response);
        }else{
            out.print("没有满足条件的 customer");
        }
    } catch (ClassNotFoundException | SQLException ex) {
```

```
        Logger.getLogger(ConditionalQueryServlet.class.getName()).log(Level.SEVERE, null, ex);
        } finally{
            //代码略
        }
    }
//代码略
}
```

showCustomer.jsp 代码目前只能写成 Java 脚本片段的形式(<% %>),代码如下:

showCustomer.jsp 代码

```
<%@page import="cn.edu.djtu.bean.Customer"%>
<%@page contentType="text/html" pageEncoding="UTF-8"%>
<!DOCTYPE html>
<html>
  <head>
    <meta http-equiv="Content-Type" content="text/html; charset=UTF-8">
    <title>显示 customer 信息</title>
  </head>
  <body>
    <%
      //1.从请求域中取出 key 为 c 的对象,
      //因为 getAttribute()方法返回值为 Object,所以需要做类型的强制转换
      Customer c = (Customer) request.getAttribute("c");
     %>
    <%-- 2.通过 Customer 的 getCustomer_id()获得 customer 的 customer_id --         %>
    <%=c.getCustomer_id()%>
    <%-- 3.通过 Customer 的 getName()获得 customer 的 name -- %>
    <%=c.getName()%>
  </body>
</html>
```

运行效果略。

7.4 使用< jsp:useBean />和< jsp:getProperty />标准动作改写案例 1

在 eg0702 的案例中,showCustomer.jsp 页面中虽然做到了功能单一,为显示数据服务,但是代码中不可避免充斥着 Java 脚本片段的形式(<% %>),以及输出表达式。那么是否有办法可以减少甚至最后让这类脚本片段完全消失呢?这就是 JSP 标准动作标记所要做的。

在项目 eg0702 上右击，选择“复制”，修改项目名称为 eg0703，如图 7.7 所示。

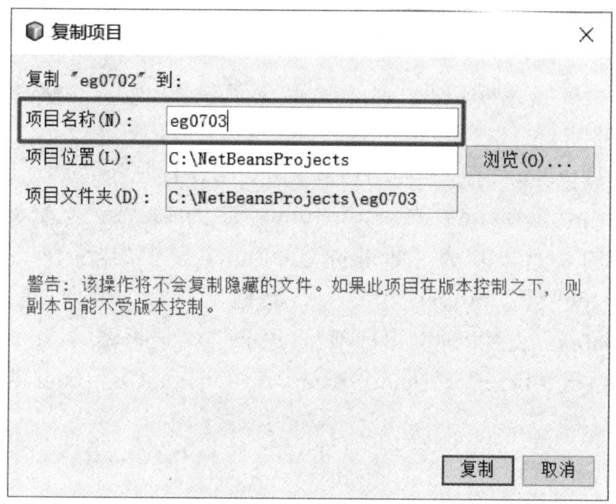

图 7.7　通过 eg0702 代码快速创建 eg0703

修改 eg0702 中 showCustomer.jsp 的代码如下：

showCustomer.jsp 代码

```
<%@page import = "cn.edu.djtu.bean.Customer" %>
<%@page contentType = "text/html" pageEncoding = "UTF-8" %>
<!DOCTYPE html>
<html>
  <head>
    <meta http-equiv = "Content-Type" content = "text/html; charset = UTF-8">
    <title>显示 customer 信息</title>
  </head>
  <body>
      <%-- //1.从请求域中取出 key 为 c 的对象
      该对象类型为 cn.edu.djtu.bean.Customer,临时引用名为 c --%>
      <jsp:useBean id = "c" class = "cn.edu.djtu.bean.Customer" scope = "request"/>
        <%--2.输出引用名为 c 的对象的属性 customer_id 的值 --%>
        <jsp:getProperty name = "c" property = "customer_id" />
        <%--3.输出引用名为 c 的对象的属性 name 的值 --%>
        <jsp:getProperty name = "c" property = "name" />
  </body>
</html>
```

可见从请求域中取出 key 为 c，类型为 cn.edu.djtu.bean.Customer 的对象，并没有使用 Java 脚本片段，而是使用了 JSP 动作标记中的<jsp:useBean />标记。使用过程中只需要设置相应的属性就可以了。

那么<jsp:useBean id=" " class=" " scope=" "/>是怎样的一个处理过程呢？

<jsp:useBean>是用来取得或者创建 JavaBean 对象的。它的处理过程是这样的，在某个范围内(由 scope 决定)，寻找引用名为某某的(由 id 决定)，某种类型的(由 class 决定)对象。如果该对象已经存在，则取得该对象；如果该对象不存在，则创建一个新对象，并存放到该范围内。其中 scope 的取值有 4 个，由小到大分别为 page、request、session 和 application。

<jsp:getProperty name="" property="" />动作标记中，name 属性取值与 useBean 标记的 id 属性对应。property 属性对应 Customer 类中的属性。整个动作标记的执行底层依赖于 Customer 类的 getter 方法。如果将 Customer 类中的 getter 方法删除的话，那么<jsp:getProperty name="" property="" />会抛出异常。例如，注释掉 eg0703 项目中 Customer 类的 getName()方法，再次运行项目，服务器端会报错，"Cannot find a method to read property 'name' in a bean of type 'cn. edu. djtu. bean. Customer'"，如图 7.8 所示。

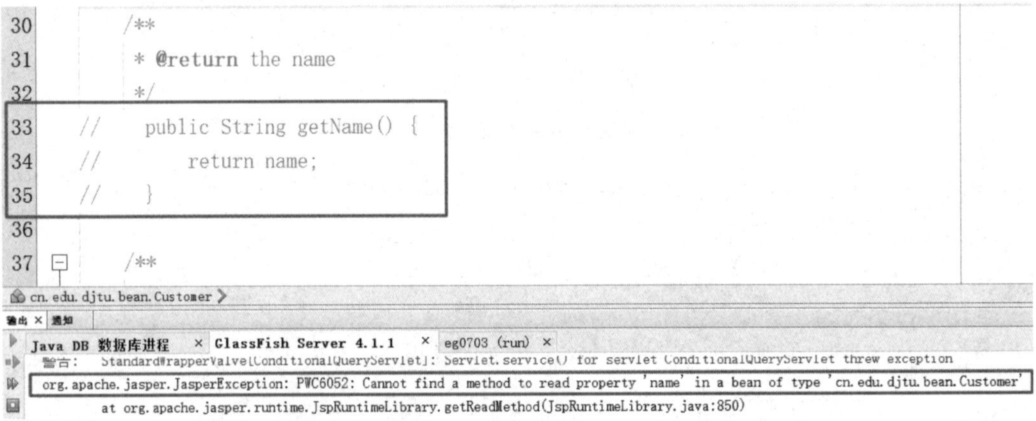

```
/**
 * @return the name
 */
//    public String getName() {
//        return name;
//    }

/**
```

```
cn.edu.djtu.bean.Customer >
输出 × 通知
Java DB 数据库进程 ×  GlassFish Server 4.1.1 ×  eg0703 (run) ×
警告:   StandardWrapperValve[ConditionalQueryServlet]: Servlet.service() for servlet ConditionalQueryServlet threw exception
org.apache.jasper.JasperException: PWC6052: Cannot find a method to read property 'name' in a bean of type 'cn.edu.djtu.bean.Customer'
        at org.apache.jasper.runtime.JspRuntimeLibrary.getReadMethod(JspRuntimeLibrary.java:850)
```

图 7.8　注释掉 getName()方法后运行 eg0703，显示异常信息

除了<jsp:useBean />和<jsp:getProperty/>以外，与 JavaBean 相关的动作标记还有<jsp:setProperty />。主要属性和用法如下。

(1) <jsp:setProperty />的 value 属性用法。

以下代码中 name 属性和 property 属性用法与<jsp:getProperty/>中的用法相同。语句的作用是将名为 c 的 JavaBean 对象的 customer_id 设置成 001。

```
<jsp:setProperty name = "c" property = "customer_id" value = "001"/>
```

(2) <jsp:setProperty />的 param 属性用法。

以下代码中 name 属性和 property 属性用法与<jsp:getProperty/>中的用法相同。语句的作用是将名为 c 的 JavaBean 对象的 customer_id 设置成变量 id 的值。

```
<jsp:setProperty name = "c" property = "customer_id" value = "id"/>
```

(3) <jsp:setProperty />的 property 属性取值为 * 的用法。

当 property 属性取值为 * 时。语句的作用是将名为 c 的 JavaBean 对象的各个属性值一一对应地设置成表单中提交的值。此时要求表单中各控件参数名与 JavaBean 中各属性名一一对应。

```
<jsp:setProperty name="c" property="*"/>
```

<jsp:useBean/><jsp:getProperty/><jsp:setProperty/>现在在开发中也不常使用了，所以不再单独作案例介绍，仅作一般性讲解。

7.5 Servlet+JSP+JavaBean 结合使用案例 2

在 eg0703 中，使用<jsp:useBean/><jsp:getProperty/>标准动作改写案例 1 后，JSP 页面中都是标签化地存在，不再有维护困难的 Java 脚本片段，那么 JSP 的标准动作是否能解决所有问题呢？请看另一个案例。

在案例 1 中，查询结果是唯一的，因为 customer_id 为 customer 表的主键。所以查询结果只可能有两种情况，要么是该 customer 存在，并且记录数是 1；要么该 customer 不存在。如果要查询的是全体 customer 的信息呢？一起来设计案例 2。

目标：学习将 Servlet+JSP+JavaBean 结合使用完成数据库查询。Servlet 负责完成业务流程控制，JSP 负责完成数据显示，JavaBean 负责完成数据封装。

工程名：eg0704。

用到的文件如表 7.2 所示。

表 7.2 eg0704 用到的文件及文件说明

文件名	说明
QueryAllCustomerServlet.java	手动创建的 Servlet 文件，用于完成数据库查询及请求转发
Customer.java	JavaBean 用于完成数据封装
showCustomer.jsp	手动创建的 JSP 文件，用于满足条件的 customer 信息输出

编程思路：

(1) QueryAllCustomerServlet 查询数据库，将每一条记录都封装成一个 customer 对象并存入列表。如果列表为空，则提示“没有满足条件的 customer”，否则将该列表存入 request 域，并将请求转发至 showCustomer.jsp。

(2) 在 showCustomer.jsp 取出该列表对象并输出列表中各 customer 的信息。

(3) 业务逻辑分析如图 7.9 所示。

打开 NetBeans IDE，新建一个名为 eg0704 的 Java Web 项目。分别新建文件名为 showCustomer.jsp 的 JSP 文件，文件名为 QueryAllCustomerServlet.java 的 Servlet 文件，文件名为 Customer.java 的文件，其中 Customer.java 文件代码与 eg0701 中的 Customer.java 完全一样。

QueryAllCustomerServlet.java 代码与之前的查询类似。

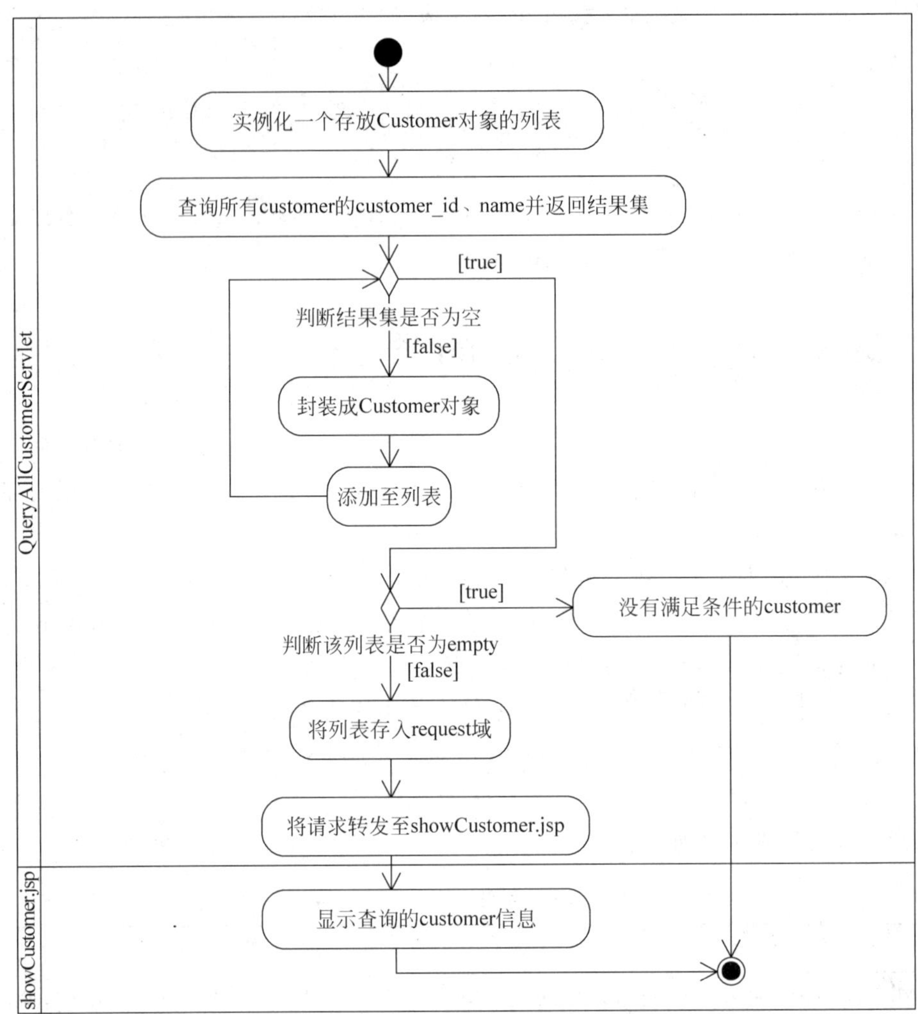

图 7.9 eg0704 业务逻辑分析

QueryAllCustomerServlet.java 代码片段

```
package cn.edu.djtu;
//代码略
@WebServlet(name = " QueryAllCustomerServlet ", urlPatterns = {"/ QueryAllCustomerServlet
"})
public class QueryAllCustomerServlet extends HttpServlet {
    protected void processRequest(HttpServletRequest request, HttpServletResponse response)
throws ServletException, IOException {
    response.setContentType("text/html;charset = UTF - 8");

    //连接数据库用到的对象
    Connection conn = null;
    PreparedStatement prst = null;
    ResultSet rs = null;
```

```
    //连接数据库用到的参数信息
    String url = "jdbc:derby://localhost:1527/sample";
    String driver = "org.apache.derby.jdbc.ClientDriver";
    String user = "app";
    String password = "app";

    //查询数据库的 SQL 语句
    String sql = "select customer_id , name from customer";

    try (PrintWriter out = response.getWriter()) {
        Class.forName(driver);
        conn = DriverManager.getConnection(url, user, password);
        prst = conn.prepareStatement(sql);
        rs = prst.executeQuery();
        List<Customer> list = new ArrayList<>();
        //只要结果集不为空,就封装对象并添加至 list
        while (rs.next()) {
          Customer c = new Customer();
          c.setCustomer_id(rs.getInt(1));
          c.setName(rs.getString(2));
          list.add(c);
        }
        //判断 list 是否为空,决定是否将请求转发至 showCustomer.jsp
        if (list.isEmpty()) {
          out.print("没有满足条件的 customer");
        } else {
          //以 key 名 list,存储了引用名为 list 的 List<Customer>对象
          request.setAttribute("list", list);
          request.getRequestDispatcher("showCustomer.jsp").forward(request, response);
        }
    } catch (ClassNotFoundException | SQLException ex) {
    Logger.getLogger(ConditionalQueryServlet.class.getName()).log(Level.SEVERE, null,
ex);
    } finally{
        //代码略
    }
  }
//代码略
}
```

showCustomer.jsp 代码目前只能采取写 Java 脚本片段的形式(<% %>)。因为 JavaBean 相关的动作标记都没有遍历 List 的能力。代码如下：

showCustomer.jsp 代码

```
<%@page import="java.util.List"%>
<%@page import="cn.edu.djtu.bean.Customer"%>
<%@page contentType="text/html" pageEncoding="UTF-8"%>
<!DOCTYPE html>
<html>
  <head>
```

```
        <meta http-equiv="Content-Type" content="text/html; charset=UTF-8">
        <title>显示 customer 信息</title>
    </head>
    <body>
        <%
List<Customer> list = (List<Customer>) request.getAttribute("list");
            for (Customer c : list) {
        %>
                customer_id:<% = c.getCustomer_id() %>
                name:<% = c.getName() %>
                <hr />
        <%
            }
        %>
    </body>
</html>
```

运行效果略。

7.6 MVC设计模式

案例1和案例2都是MVC设计模式的一种具体实现。那么什么是MVC呢?

所谓MVC(Model View Controller),指的就是模型-视图-控制器。MVC是一种成熟的设计模式,它并不是Java平台独有的,它将一个应用程序在逻辑上视为三层。负责用户界面的叫作视图层(View),负责程序流程控制的叫作控制器(Controller),负责对数据进行建模的叫作模型(Model)。

在Java Web应用程序开发中,曾经产生过模式一(JSP+JavaBean),以及模式二(Servlet+JSP+JavaBean)。模式二就是MVC设计模式的一种实现。在模式二中,JavaBean用来对数据建模,完成了模型层的工作;JSP用来与用户交互,完成了视图层的工作;Servlet用来控制程序的流转,完成控制器层的工作。

采用分层的设计思想后,程序变得易于扩展和维护,因为每一层都可以由不同的技术实现。层与层之间的耦合程度变小,某一层的变化不会影响其他层。例如在eg0702和eg0703中,showCustomer.jsp的代码不同,但是实现的功能完全相同。在第8章将采用另一种实现方式来修改eg0704中showCustomer.jsp的代码实现,同样可以看到,视图层的改变不会影响其他两层。

7.7 关于JSP动作标记的思考

在7.4节最后,尝试注释掉Customer类的getName()方法,发现<jsp:getProperty />动作标记失效了,验证了<jsp:getProperty />动作标记实际上调用的就是getName()方法。透过这个例子,可以窥视一下JSP动作标记背后蕴含的思想。

任何技术的产生和发展都有其特定的背景,JSP动作标记的思想是:将具体的业务操作封装在类中，在 JSP 页面上尽量减少业务逻辑的处理，而代之以实例方法的调用。这种调用以标签的形式出现，与 HTML 标签形式类似，这样做页面代码风格既和谐又统一。通过设置标签的属性进行相关的具体操作。

这种思想的缺陷是现有的 JSP 标准动作标签无法覆盖用户的全部需求。所以会看到现在的企业级开发中，一般都是使用框架开发，使用某个框架，就使用该框架自定义的标签。框架中的标签可以看作是 JSP 动作标记的延伸。也有的公司干脆自定义标签库，用公司专用的架构。

有鉴于此，对于 JavaBean 相关的动作标记以及 JSP 其他的动作标记，本书就不再做过多的介绍和讲解了。

7.8 本章回顾

本章首先介绍了 JavaBean 的定义及分类，接着介绍了 JavaBean 的语法规则，然后将 Servlet＋JSP＋JavaBean 三者结合完成了一个案例；接着使用< jsp: useBean/>和< jsp: getProperty/>标准动作改写案例。通过案例可以看到，在 JSP 页面中通过使用 JavaBean 相关的标准动作标签，一定程度上能够减少页面上的 Java 脚本，增强可维护性，同时可以提高开发效率。但是，JavaBean 并不是解决问题的万能钥匙，它的使用并不能在页面上根除 Java 脚本，Servlet＋JSP＋JavaBean 结合的案例 2 体现了这一点。最后总结了 MVC 设计模式以及对 JSP 动作标记的思考。本章小结如图 7.10 所示。

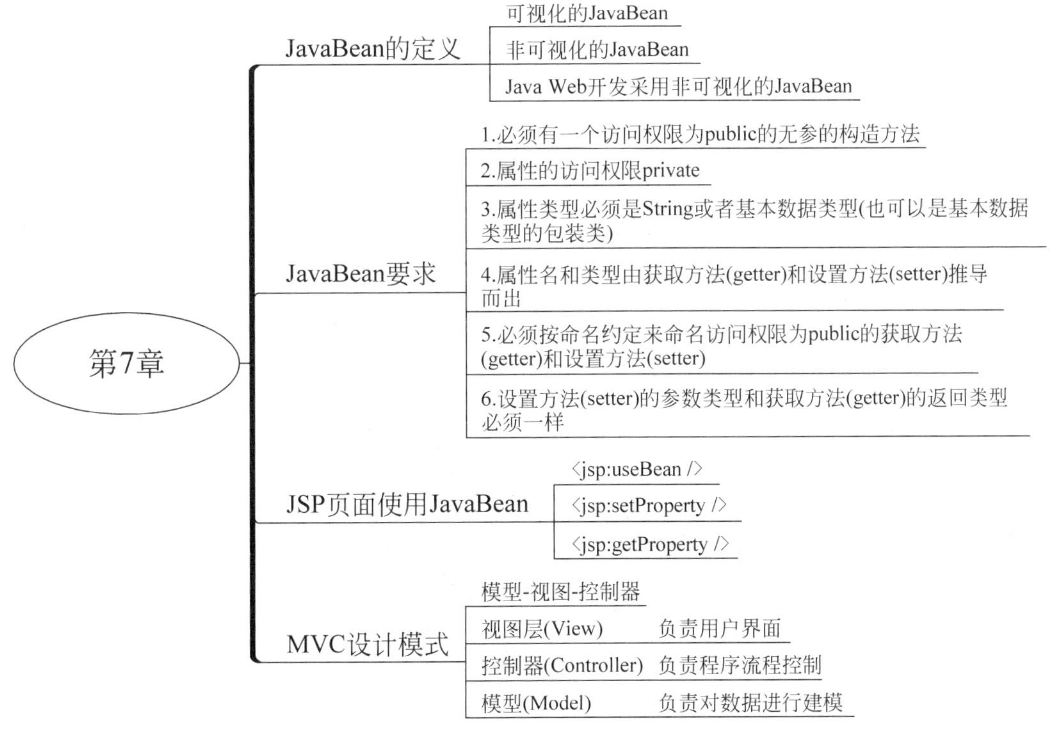

图 7.10 第 7 章内容结构图

7.9 课后习题

1. JavaBean 是什么？分成几类？Java Web 开发中主要使用哪一类的 JavaBean？

2. JavaBean 的语法规则有哪些？

3. 与 JavaBean 相关的动作标记有哪些？请结合代码说明。

4. 如果想以 property="*"的形式设置 JavaBean 对象，对 JavaBean 和提交信息的页面有什么要求？

5. 简述 MVC 设计模式。

第8章 使用EL与JSTL

学习目标：

通过本章的学习，你应该：

- 了解表达式语言(EL)的优点
- 掌握EL的基本语法
- 掌握EL运算符的应用
- 掌握如何使用EL对表达式进行条件求值
- 掌握EL中与作用域相关的4个隐含对象
- 掌握EL中“.”操作符与“[ ]”操作符的用法
- 掌握如何使用EL访问JavaBean的属性
- 掌握JSTL中流程控制标签的用法
- 掌握JSTL中<c:forEach></c:forEach>标签的用法

从7.5节的案例2(eg0704)可以看到，即使引入了JavaBean相关的JSP标准动作，也无法做到JSP页面没有Java脚本片段。那么有没有办法更彻底地去除JSP页面中的Java脚本片段呢？本章的EL与JSTL结合就可以做到这一点。

在本章的学习中，并不要求掌握EL和JSTL的每一个细节。EL在当前的企业级开发中依然有使用，JSTL的应用就很少见了。因为可替代JSTL的框架很多，也比JSTL要完备一些。所以在学习中重点学习EL，JSTL要窥一斑而见全豹，掌握思想即可。下面先学习EL。

8.1 EL(表达式语言)的使用

1. EL概述与基本语法

EL(Expression Language)叫作表达式语言，它并不是JSP中一开始就有的，而是JSP 2.0新增的技术规范。如果是维护旧版本的系统，或者对旧的系统升级，那么使用EL还需要小心设置。本章讨论都是假定默认支持EL的。至于如何和旧的系统兼容，需要大家遇到问题的时候查阅相关资料。

EL的设计主要目的是简化页面输出，它能够完美替代传统JSP中的输出表达式<%=%>，而且还提供了额外的功能，例如能以更自然的方式输出JavaBean的属性信息等。

它的语法非常简单，以 $ 符号开始，代码写在{ }之间就可以了。大致的结构是这样的。后面会通过例子详细地学习 EL。

```
${某个经计算有值的表达式}
```

2. EL 输出常量和变量的值

通过 EL 输出常量或者变量的值，用法与输出表达式<%=%>基本相同，不一样的是，当变量是 null 的时候，使用输出表达式<%=%>可能会抛出空指针异常，而 EL 不会抛出异常。此外，EL 还做了一部分变量类型自动转换的工作。

先看输出常量，实现 EL 版 Hello World！的代码如下：

```
<%@page contentType="text/html" pageEncoding="UTF-8"%>
<!DOCTYPE html>
<html>
    <head>
        <meta http-equiv="Content-Type" content="text/html; charset=UTF-8">
        <title>JSP Page</title>
    </head>
    <body>
        <h1>${"Hello World!"}</h1>
    </body>
</html>
```

再看输出变量，同样是输出“Hello World!”，代码如下，运行效果略(对于这段代码，暂时只需要简单了解，可以看到能够通过变量名的形式输出就可以了，详见 8.1.6 节)。

```
<%@page contentType="text/html" pageEncoding="UTF-8"%>
<!DOCTYPE html>
<html>
    <head>
        <meta http-equiv="Content-Type" content="text/html; charset=UTF-8">
        <title>JSP Page</title>
    </head>
    <body>
        <%
            request.setAttribute("msg", 1234);
        %>
        ${msg}
    </body>
</html>
```

与输出表达式<%=%>一样，EL 既可以直接写在 HTML 的标签内部，也可以写在某个标签的属性中，例如：

```
<%
    request.setAttribute("url", "index.jsp");
%>
<a href = "${url}">这是一个超级链接,链接到首页</a>
```

以上代码最终在客户端输出时,源代码如下:

```
<a href = "index.jsp">这是一个超级链接,链接到首页</a>
```

3. EL 的运算符

EL 提供了比较丰富的运算符。包括算术运算符、逻辑运算符、关系运算符,还有空运算符(empty)。

1) 算术运算符

算术运算符有加法(+)、减法(−)、乘法(*)、除法(/或者 div)、取余(%或者 mod),如表 8.1 所示。用法与其他语言大体相同,但是其中有几个地方值得注意:

(1) 对于加法(+)、减法(−)运算,如果参与运算的任意操作数为字符串,运行时都将被自动解析成数值型,如果该操作数不能被解析成数值,则会抛出 java.lang.NumberFormatException 的异常;如果参与运算的任意一个操作数为 BigInteger 或 BigDecimal 类型,则与 Java 中的要求一样,需要做类型转换,否则会损失精度。转换后等价于调用相应的 add 和 subtract 方法。

(2) 表 8.1 中 EL 表达式 ${3/0},在运算中该语句并不会抛出异常,而是显示为 Infinity(无穷),如果是<%=3/0%>,则会抛出 java.lang.ArithmeticException:/by zero 的异常。

(3) 表 8.1 中最后一行是一个三目运算符,与 C 语言中的三目运算符用法一样。

表 8.1　EL 中的算术运算符

EL 表达式	结　果
${1}	1
${1+2}	3
${1.2+2.3}	3.5
${1.2E4+1.4}	12001.4
${−4 − 2}	−6
${21 * 2}	42
${3/4}或${3 div 4}	0.75
${3/0}	infinity
${10%4}或${10 mod 4}	2
${(1==2) ? 3 : 4}	4

2) 逻辑运算符

逻辑运算符有与(&& 或者 and)、或(|| 或者 or)、非(not)三种。与 Java 语言中一样,如表 8.2 所示。

表 8.2 EL 中的逻辑运算符

<table>
<tr><th>EL 表达式</th><th>结　果</th></tr>
<tr><td>${true and false}或${true && false}</td><td>false</td></tr>
<tr><td>${true or false}或${true || false}</td><td>true</td></tr>
<tr><td>${not true}或${! true}</td><td>false</td></tr>
</table>

3) 关系运算符

关系运算符有小于(<或 lt)、大于(>或 gt)、大于等于(>=或 ge)、小于等于(<=或 le)、等于(==或 eq)、不等于(! =或 ne),如表 8.3 所示。

(1) 等于(==或 eq),总体上与 Java 中的 equals 方法类似,如果两个操作数是同一个对象,则返回 true。如果两个操作数为数值,则与 Java 中的==运算符一样。如果任意一个操作数为 null,则返回 false。如果任意一个操作数为 BigInteger 或 BigDecimal 类型,则与 Java 中的要求一样,需要做类型转换,否则会损失精度,转换后等价于用 compareTo 方法比较操作数。

(2) 不等于(! =或 ne)处理时与等于(==或 eq)运算符类似。只是逻辑正好相反。

(3) 小于(<或 lt)、大于(>或 gt)、大于等于(>=或 ge)、小于等于(<=或 le)处理时与等于(==或 eq)运算符类似。另外,当操作数为字符串时,关系运算将进行字面比较,如表中的倒数第 3 行和倒数第 2 行。

表 8.3 EL 中的逻辑运算符

EL 表达式	结果
${1 < 2}或${1 lt 2}	true
${1 > (4/2)}或${1 gt (4/2)}	false
${4.0 >= 3}或${4.0 ge 3}	true
${4 <= 3}或${4 le 3}	false
${100.0 == 100}或${100.0 eq 100}	true
${(10 * 10) ! = 100}或${(10 * 10) ne 100}	false
${'a'< 'b'}	true
${"hip" > "hit"}	false
${'4' > 3}	true

4) 空运算符

空运算符(empty)用来判断参数是不是 null、空字符串、空数组、空 Map 或者空集合,如果是则返回 true,否则返回 false,如表 8.4 所示。

表 8.4 EL 中的空运算符

EL 表达式	结果
${empty null}	true
${empty ""}	true

注意:以上有关运算符的例子主要来自 Tomcat 中关于 EL 的例子,感兴趣的可以自行查阅。在 Tomcat 安装目录下的\webapps\examples\jsp\jsp2\el 内。在表格中没有给出有

关 BigInteger 或 BigDecimal 类型相关的运算符例子，这是因为 EL 的主要作用在于简化输出，提供对现有对象的简洁访问，对于复杂的应用和业务逻辑，以及相应的数据处理，不应该依赖于 EL，而应该放在 Java 类中进行处理。况且在 Java 类中处理这两种类型也要经过特别的处理。

以上内容侧重于验证性，能理解即可，不需要每个例子都编写。特别地，empty 运算符是其他语言里面没有的，而后面案例会用到，所以注意体会其用法。

4. EL 中与作用域相关的隐含对象

前面 3 个小节展示了 EL 与 JSP 输出表达式<%＝%>类似功能的部分。除了一般的输出任务，EL 还提供了更为简洁的语法形式，以便访问或输出某个域中对象的属性。

在第 6 章 JSP 基础中，曾经介绍过 JSP 中的隐含对象，在第 7 章 JSP 与 JavaBean 中介绍了 UseBean 的 scope 属性(Scope 属性的取值是固定的，有 4 个值可供选择 page、request、session、application，可以访问这 4 个作用域中的某个对象)。其实在 EL 中也提供了一系列隐含对象。本书中只详细介绍 EL 中的 4 个与作用域相关的隐含对象，这四个对象都是映射类型(Map)，可以通过 key-value 的形式访问。全部的隐含对象如表 8.5 所示。

表 8.5 EL 中的隐含对象

隐含对象	描述
pageScope	page 作用域
requestScope	request 作用域
sessionScope	session 作用域
applicationScope	application 作用域
param	request 对象的参数，字符串
paramValues	request 对象的参数，字符串集合
header	HTTP 信息头，字符串
headerValues	HTTP 信息头，字符串集合
initParam	上下文初始化参数
cookie	cookie 值
pageContext	当前页面的 pageContext

为什么只重点介绍这 4 个隐含变量，而不是其他的呢？这是因为这 4 个对象与要做的共享数据对象这件事密切相关。在第 7 章结束的时候，将 JSP、Servlet 和 JavaBean 结合起来完成了两个例子。这两个例子的操作流程抽象出来是一样的：用户向 Servlet 发送请求；Servlet 进行处理，将处理结果封装成对象存储在某个作用域里，然后通过请求的转发器将控制权移交给 JSP 页面；在页面上采用某种机制输出相关对象的信息。在 Servlet 中通过 setAttribute 方法存储数据，而在页面上采用了两种不同的方法来取数据，如图 8.1 所示。

那么如何利用 EL 来取出作用域中的对象呢？EL 提供了两个操作符，即“.”操作符与“[ ]”操作符。

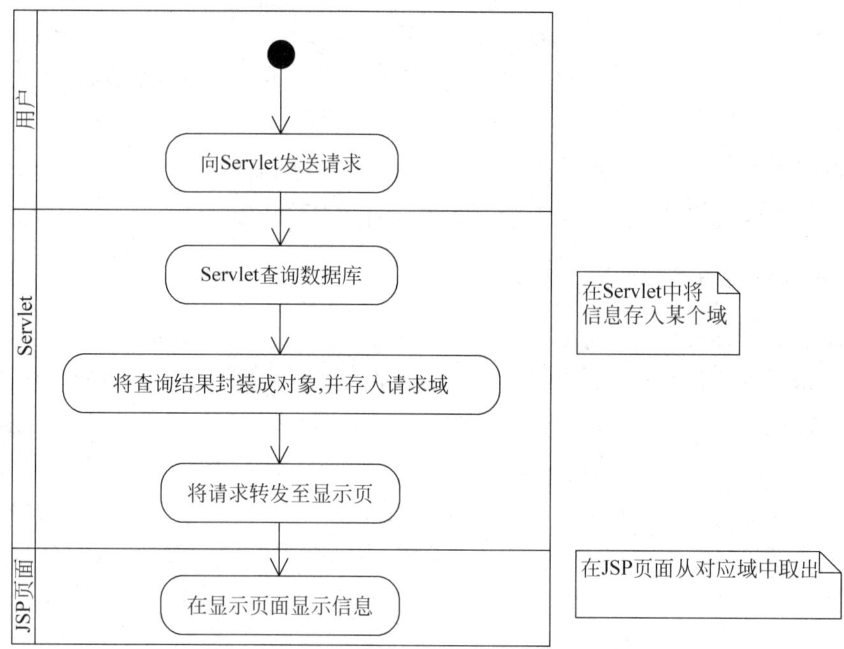

图 8.1 业务处理流程活动图

5. EL 中的"."操作符

目标：学习"."操作符的使用。

工程名：eg0801。

用到的文件如表 8.6 所示。

表 8.6 eg0801 用到的文件及文件说明

文件名	说明
Customer.java	与第 7 章中的 Customer.java 代码完全一样,手动创建的 JavaBean,封装了 Customer 的信息,有 customer_id、name 两个字段及相应的 setter/getter 方法
EL1.jsp	手动创建的 JSP 文件,使用"."操作符输出 Map 类型对象的信息
EL2.jsp	手动创建的 JSP 文件,使用"."操作符输出 JavaBean 类型对象的信息

打开 NetBeans IDE,新建一个名为 eg0801 的 Java Web 项目,具体过程略。新建名为 Customer.java 的 JavaBean(也可以直接从第 7 章项目中复制,再粘贴到 eg0801 的源包目录下);再新建名为 EL1.jsp、EL2.jsp 的 JSP 页面,具体过程略。

"."操作符的操作对象主要是映射类型(Map)和 JavaBean 类型,对于映射类型(Map),可以通过键值形式访问到。对于 JavaBean 类型,可以通过属性名的形式访问。EL1.jsp 演示了如何使用"."操作符输出 Map 类型对象的信息;EL2.jsp 演示了如何使用"."操作符输出 JavaBean 类型对象的信息。所操作的对象都存储在 request 域中。

EL1.jsp 的代码如下：

EL1.jsp

```
<%@page import="java.util.HashMap" %>
<%@page import="java.util.Map" %>
<%@page contentType="text/html" pageEncoding="UTF-8" %>
<!DOCTYPE html>
<html>
    <head>
        <meta http-equiv="Content-Type" content="text/html; charset=UTF-8">
        <title>EL1</title>
    </head>
    <body>
        <h1>使用"."操作符输出 Map 类型对象的信息</h1>
        <%
            Map<String, String> fruitMap = new HashMap ();
            fruitMap.put("f1", "香蕉");
            fruitMap.put("f2", "苹果");
            fruitMap.put("f3", "哈密瓜");
            request.setAttribute("fmap", fruitMap);
         %>
        第 1 种水果: ${requestScope.fmap.f1}<hr />
        第 2 种水果: ${requestScope.fmap.f2}<hr />
        第 3 种水果: ${requestScope.fmap.f3}<hr />
        全部的水果: ${requestScope.fmap}
    </body>
</html>
```

运行效果如图 8.2 所示，注意看最后一行语句的输出，当没有指明键值时，与调用对象的 toString()方法的效果一样。

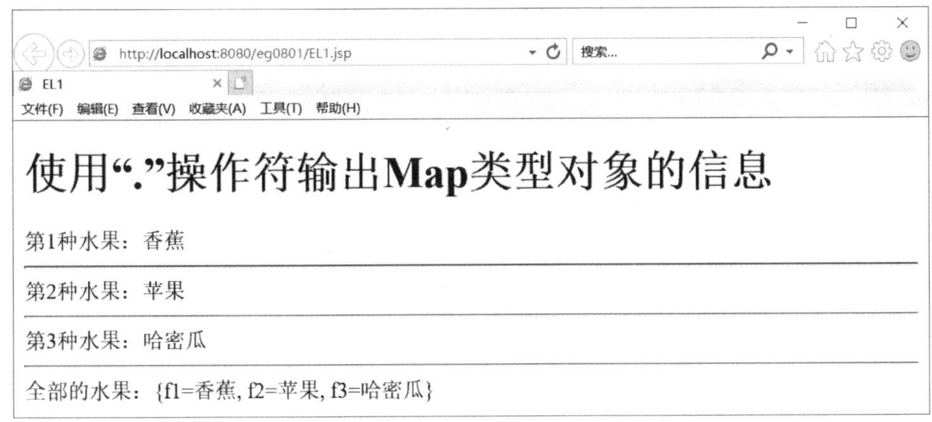

图 8.2 使用"."操作符输出 Map 类型对象的信息

此外需要注意的是：如果在当前环境下，代码写成：

```
Map<String, String> fruitMap = new HashMap<>();
```

那么程序会报错：

```
PWC6199: Generated servlet error:
- source 1.5 中不支持 diamond 运算符
(请使用 -source 7 或更高版本以启用 diamond 运算符)
```

这是运行环境版本的设置问题。GlassFish 4.0 默认使用 JDK 1.5 将 JSP 转换成 Servlet，不支持类型泛型的类型推断。Tomcat 8.0 也有类似的问题，Tomcat 8.0 默认使用 JDK 1.6。本次练习可以暂不修改环境，只调整一下代码即可。

```
Map<String, String> fruitMap = new HashMap();
```

编写 EL2.jsp 页面的代码，这里使用了 JSP 的标准动作来实例化、初始化对象，然后存储在 request 域中。

EL2.jsp

```
<%@page contentType="text/html" pageEncoding="UTF-8"%>
<!DOCTYPE html>
<html>
    <head>
    <meta http-equiv="Content-Type" content="text/html; charset=UTF-8">
    <title>EL2</title>
    </head>
    <body>
        <h1>使用"."操作符输出 JavaBean 类型对象的信息</h1>
<jsp:useBean id="customer" class="cn.edu.djtu.bean.Customer" scope="request"/>
<jsp:setProperty name="customer" property="customer_id" value="1"/>
<jsp:setProperty name="customer" property="name" value="青椒麦客"/>
        ${requestScope.customer.customer_id}<hr />
        ${requestScope.customer.name}<hr />
        输出该对象?${requestScope.customer}
    </body>
</html>
```

运行效果如图 8.3 所示。最后一行语句的输出，当没有指明属性时，与调用对象的 toString()方法的效果一样。

6. EL 中的“[ ]”操作符

如果从作用域中取出的对象不是映射类型(Map)或者不是 JavaBean 的类型，那么“.”操作符就无能为力了，这时候就必须使用“[ ]”操作符。

目标：学习 EL 中“[ ]”操作符的使用。

工程名：使用已经创建的 eg0801。

用到的文件如表 8.7 所示。

图 8.3　使用“.”操作符输出 JavaBean 类型对象的信息

表 8.7　学习 EL 中“[ ]”操作符的使用所用到的文件

文件名	说　　明
EL3.jsp	手动创建的 JSP 文件，使用“[ ]”操作符输出 Map 类型对象的信息
EL4.jsp	手动创建的 JSP 文件，使用“[ ]”操作符输出 JavaBean 类型对象的信息
EL5.jsp	手动创建的 JSP 文件，使用“[ ]”操作符输出数组类型对象的信息
EL6.jsp	手动创建的 JSP 文件，使用“[ ]”操作符输出 List 类型对象的信息

在 eg0801 的 Java Web 项目中新建名为 EL3.jsp、EL4.jsp、EL5.jsp、EL6.jsp 的 JSP 页面，具体过程略。

“[ ]”操作符更为灵活，它既可以像“.”操作符一样处理映射类型(Map)(见 EL3.jsp)、JavaBean 的类型(见 EL4.jsp)，又能操作数组类型对象(见 EL5.jsp)、列表类型(List)(见 EL6.jsp)。

编写 EL3.jsp 代码如下，使用“[ ]”操作符输出 Map 类型对象的信息时，参数同样为键值。

EL3.jsp

```
<%@page import="java.util.HashMap"%>
<%@page import="java.util.Map"%>
<%@page contentType="text/html" pageEncoding="UTF-8"%>
<!DOCTYPE html>
<html>
    <head>
        <meta http-equiv="Content-Type" content="text/html; charset=UTF-8">
        <title>EL3</title>
    </head>
    <body>
        <h1>使用"."操作符输出 Map 类型对象的信息</h1>
        <%
            Map<String, String> fruitMap = new HashMap();
            fruitMap.put("f1", "香蕉");
            fruitMap.put("f2", "苹果");
```

```
            fruitMap.put("f3", "哈密瓜");
            request.setAttribute("fmap", fruitMap);
        %>
        第 1 种水果: ${requestScope.fmap["f1"]}<hr />
        第 2 种水果: ${requestScope.fmap["f2"]}<hr />
        第 3 种水果: ${requestScope.fmap["f3"]}<hr />
        全部的水果: ${requestScope.fmap}
    </body>
</html>
```

运行效果如图 8.4 所示。

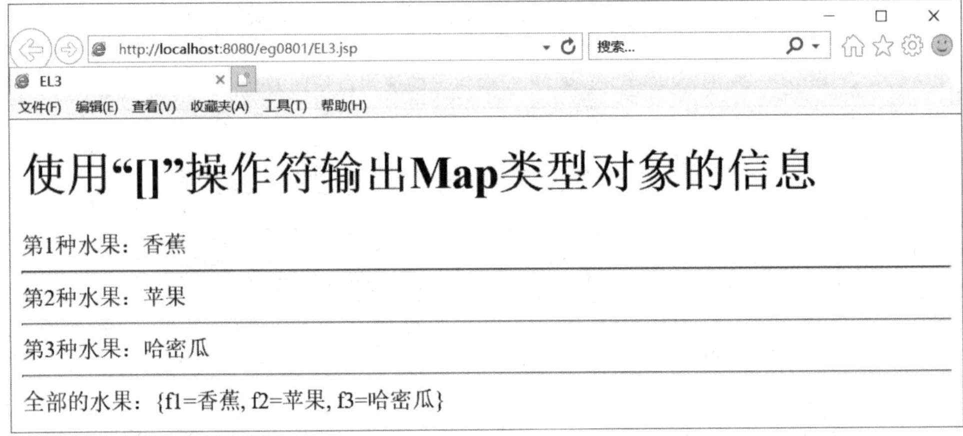

图 8.4 使用“[ ]”操作符输出 Map 类型对象的信息

编写 EL4.jsp 代码如下，使用“[ ]”操作符输出 JavaBean 类型对象的信息时，参数为属性名。

EL4.jsp

```
<%@page contentType="text/html" pageEncoding="UTF-8"%>
<!DOCTYPE html>
<html>
    <head>
    <meta http-equiv="Content-Type" content="text/html; charset=UTF-8">
    <title>EL4</title>
    </head>
    <body>
        <h1>使用"[]"操作符输出 JavaBean 类型对象的信息</h1>
        <jsp:useBean id="customer" class="cn.edu.djtu.bean.Customer" scope="request"/>
        <jsp:setProperty name="customer" property="customer_id" value="1"/>
        <jsp:setProperty name="customer" property="name" value="青椒麦客"/>
        ${requestScope.customer["customer_id"]}<hr />
        ${requestScope.customer["name"]}<hr />
        输出该对象? ${requestScope.customer}
    </body>
</html>
```

运行效果如图 8.5 所示。

使用“[]”操作符输出JavaBean类型对象的信息

1

青椒麦客

输出该对象? cn.edu.djtu.bean.Customer@5c787898

图 8.5 使用“[ ]”操作符输出 JavaBean 类型对象的信息

编写 EL5.jsp 代码如下：

EL5.jsp

```
<%@page contentType="text/html" pageEncoding="UTF-8"%>
<!DOCTYPE html>
<html>
    <head>
        <meta http-equiv="Content-Type" content="text/html; charset=UTF-8">
        <title>EL5</title>
    </head>
    <body>
        <h1>使用"[]"操作符输出数组类型对象的信息</h1><%
            String[] fruits = {"香蕉","苹果","哈密瓜"};
            request.setAttribute("farray", fruits);
        %>
        第 1 种水果: ${requestScope.farray[0]}<hr />
        第 2 种水果: ${requestScope.farray[1]}<hr />
        第 3 种水果: ${requestScope.farray["2"]}<hr />
        全部的水果?: ${requestScope.farray}
    </body>
</html>
```

运行效果如图 8.6 所示。

使用“[ ]”操作符输出数组类型对象的信息时，参数为索引值。下标从 0 开始，乍一看“[ ]”操作符似乎就是数组访问操作符，但其实不是，因为“数组访问操作符[ ]”中一定是整型的下标值，请看第 3 种水果部分的代码。

```
第 3 种水果: ${requestScope.farray["2"]}<hr/>
```

即使索引值的部分写成字符串形式的"2"，依然可以顺利运行并且达到设计意图。这足以证明“[ ]”操作符与数组访问操作符是不同的。这个字符串形式的参数会被 EL 自动转

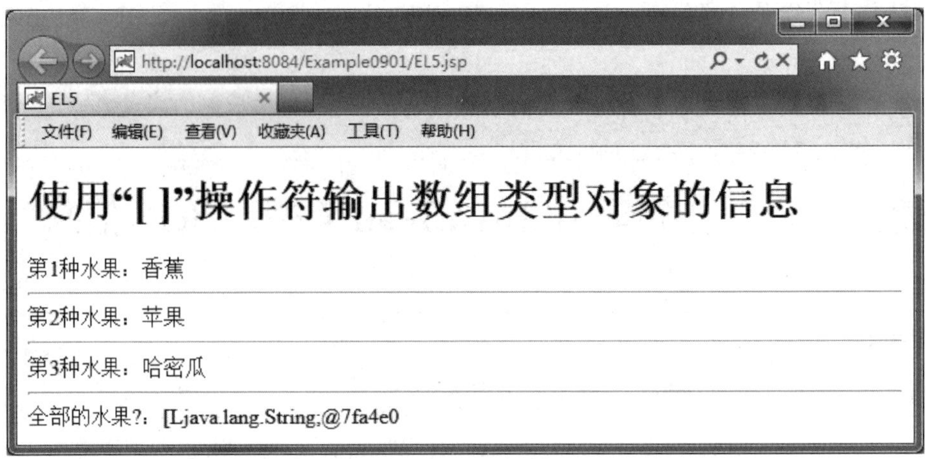

图 8.6 使用“[ ]”操作符输出数组类型对象的信息

换成 int 型，如果无法转换，例如输入的是"a"，则抛出 java.lang.NumberFormatException：For input string："a"的异常。

使用“[ ]”操作符输出 List 类型对象的信息与输出数组类型对象的信息一样。编写 EL6.jsp 代码如下：

EL6.jsp

```
<%@page import="java.util.List"%>
<%@page import="java.util.ArrayList"%>
<%@page contentType="text/html" pageEncoding="UTF-8"%>
<!DOCTYPE html>
<html>
    <head>
        <meta http-equiv="Content-Type" content="text/html; charset=UTF-8">
        <title>EL6</title>
    </head>
    <body>
        <h1>使用"[ ]"操作符输出 List 类型对象的信息</h1>
        <%
            List<String> fruitList = new ArrayList();
            fruitList.add("香蕉");
            fruitList.add("苹果");
            fruitList.add("哈密瓜");
            request.setAttribute("flist", fruitList);
        %>
        第 1 种水果: ${requestScope.flist[0]}<hr />
        第 2 种水果: ${requestScope.flist[1]}<hr />
        第 3 种水果: ${requestScope.flist[2]}<hr />
        全部的水果: ${requestScope.flist}
    </body>
</html>
```

运行效果如图 8.7 所示。

http://localhost:8080/eg0801/EL6.jsp

EL6

文件(F) 编辑(E) 查看(V) 收藏夹(A) 工具(T) 帮助(H)

使用“[ ]”操作符输出List类型对象的信息

第1种水果：香蕉

第2种水果：苹果

第3种水果：哈密瓜

全部的水果：[香蕉, 苹果, 哈密瓜]

图 8.7 使用“[ ]”操作符输出 List 类型对象的信息

从以上的例子可以看出，“[ ]”操作符像是增强版的“.”操作符，除了可操作的对象类型更为丰富以外，“[ ]”操作符的优点还有很多，但是考虑到在本书的讨论范围内，多数是操作 JavaBean 类型的对象，所以就不做过多的讲解了，如果在工程中遇到需要使用“[ ]”操作符的地方，请查阅其他资料。

7. EL 如何在作用域中查找对象

从“5. EL 中的‘·’操作符”开始，已经在例子中尝试访问了 4 个与作用域相关的隐含对象中的 requestScope，那么其他的 3 个又该如何访问呢？下面再通过一个例子来总结一下。

初步的设想是这样的：先通过 Java 脚本在“当前页(pageContext)”“请求(request)”“会话(session)”“应用(application)”四个作用域里各自封装一个 key 名是"myKey"的对象，但是对应的 value 不同，如下所示：

```
<%
    pageContext.setAttribute("myKey", "pageScopeValue");
    request.setAttribute("myKey", "requestScopeValue");
    session.setAttribute("myKey", "sessionScopeValue");
    application.setAttribute("myKey", "applicationScopeValue");
%>
```

然后再用 EL 将各个域里面 key 值为"myKey"的对象取出，如下所示：

```
输出 pageScope 中键值"myKey"对应的对象 ${pageScope.myKey}<hr/>
输出 requestScope 中键值"myKey"对应的对象 ${requestScope.myKey}<hr/>
输出 sessionScope 中键值"myKey"对应的对象 ${sessionScope.myKey}<hr/>
输出 application 中键值"myKey"对应的对象 ${applicationScope.myKey}<hr/>
```

最后再看看更为奇妙的事情，例如，它可能是这样（其中为了在页面输出形如“${myKey}”的字符串，使用了转义）：

```
\${myKey}输出的是这个：${myKey}<hr/>
```

首先在 eg0801 中新建一个名为 EL7.jsp 的页面，完整代码如下：

EL7.jsp

```
<%@page contentType="text/html" pageEncoding="UTF-8"%>
<!DOCTYPE html>
<html>
    <head>
        <meta http-equiv="Content-Type" content="text/html; charset=UTF-8">
        <title>EL7</title>
    </head>
    <body>
        <h1>EL 如何在作用域中查找对象</h1>
        <%
            pageContext.setAttribute("myKey", "pageScopeValue");
            request.setAttribute("myKey", "requestScopeValue");
            session.setAttribute("myKey", "sessionScopeValue");
            application.setAttribute("myKey", "applicationScopeValue");
        %>
        输出 pageScope 中键值"myKey"对应的对象 ${pageScope.myKey}<hr/>
        输出 requestScope 中键值"myKey"对应的对象 ${requestScope.myKey}<hr/>
        输出 sessionScope 中键值"myKey"对应的对象 ${sessionScope.myKey}<hr/>
        输出 application 中键值"myKey"对应的对象 ${applicationScope.myKey}<hr/>
        \${myKey}输出的是这个：${myKey}<hr/>
</body>
</html>
```

运行结果如图 8.8 所示。

图 8.8　EL 如何在作用域中查找对象

从图中可以看出，尽管各个对象的键值相同，但是由于存储在不同域里，所以从各个域中取出时，输出各不相同。

最后这条语句的输出可能让人迷惑，并没有指定哪个域，它是如何将 myKey 对应的 value 取出来的呢？

```
${myKey}
```

道理很简单，当没有明确指明对象的作用域时，EL 将根据作用域范围按照从小到大的顺序搜索满足条件的 key 值，找到后即输出。也就是按照 pageScope、requestScope、sessionScope、applicationScope 这样的顺序查找，将查找到的第一个满足条件的对象返回。刚才的例子中，输出是“pageScope.myKey”，说明找到的是 pageScope 中的名为“myKey”的对象，尽管其他 3 个作用域中也有同名的对象，但也不再继续寻找。

事实上在 Servlet 中存储对象时，一般都是唯一的 key 值，所以多数情况下都不需要指明作用域，简写成 ${myKey}的形式就可以了，如果特殊情况下有需要，那么就像例子里面一样，指明相应的作用域即可。

接下来练习刚刚学过的 EL 知识。

8. 使用 EL 改写案例 1

在第 7 章的 eg0703 中，在 showCustomer.jsp 页面上通过 JSP 标准动作访问了请求域中的 JavaBean 对象，复制 eg0703，在弹出的窗口将项目名称改为 eg0802。这样就拥有了和 eg0703 完全相同的代码。改写 showCustomer.jsp 代码如下：

showCustomer.jsp

```
<%@page contentType="text/html" pageEncoding="UTF-8"%>
<!DOCTYPE html>
<html>
    <head>
        <meta http-equiv="Content-Type" content="text/html; charset=UTF-8">
        <title>显示 customer 信息</title>
    </head>
    <body>
        customer_id: ${c.customer_id}<hr/>
        name: ${c.name}
    </body>
</html>
```

可以看到，使用 EL 更加直观简洁。可见 EL 能够很好地完成输出任务。不但能顺利取出对象，还可以继续用“.”运算符取出该对象的属性，非常方便。形式如下：

作用域.对象键值.属性名。

例子中省略了作用域，直接用 c.customer_id 的形式输出了 customer_id 信息。

是否可以用 EL 来改写一下 eg0704 呢？这样不行，因为 eg0704 中涉及遍历（循环）操作，EL 没有这样的机制完成这个任务，这时候就要考虑其他的技术，JSTL 是其中的一个解决方案。

8.2 JSTL(JSP 标准标签库)的使用

1. JSTL 概述

JSTL(JavaServer Pages Standard Tag Library)中文意思是 JSP 标准标签库。但是可能令人费解的是虽然叫作标准标签库，它却不能像 EL 和其他的标准动作标签一样直接使用，而是需要先导入相应的 jar 文件。

JSTL 最初由 Sun 公司提供，封装常见的 JSP 应用程序的核心功能。它提供了一套统一的、标准的标签组。这种标准化设计，使得应用程序可以轻松地部署在任何支持 JSTL 的 JSP 容器内，而无须定制开发。考虑到 JSP 页面主要用于输入输出，而不应该有过多的业务逻辑操作，这样的标签库已经足够应付大多数应用了。

JSTL 有 5 个不同功能的标签库，分别对应着不同的应用场合，例如：

- core：主要负责核心功能实现，包括变量支持、流程控制、URL 管理等。
- XML：XML 文档相关，XML 核心操作、XML 流程控制、XML 转换。
- I18N：主要负责国际化、格式化(日期、时间、数字的格式化)。
- SQL：数据库操作相关。
- Functions：在 JSTL 中定义的标准 EL 函数集，主要是对字符串的操作。

本书仅以其中核心库中的几个标签为例，做一个简要的介绍，掌握其原理。

2. JSTL 的版本及 taglib 声明

JSTL 目前有 3 个版本：JSTL 1.0、JSTL 1.1、JSTL 1.2。JSTL 1.0 和 JSTL 1.1 包含 jstl.jar 和 standard.jar 这两个 jar 包，在 J2EE4 的版本中是需要单独引用这两个 jar 包的。

JSTL 1.2 在 J2EE5 里的 jar 包是 jstl-1.2.jar，在 J2EE6 里是 jstl-imp.jar。本书使用的版本是 JSTL1.2。

在 Web 项目中使用 JSTL，需要先下载相关的 jar 包，然后导入到工程中。而在 NetBeans 中使用 JSTL 比较简单，下面举例说明。

打开 NetBeans IDE，新建一个名为 Example0803 的 Java Web 项目，具体过程略。选中“库”，右击，在弹出的菜单中选择“添加库”，在弹出的窗口中找到 JSTL 1.2.1，选择“添加库”，添加完成后，可以看到“库”下面多了两个文件，如图 8.9 所示。

因为 GlassFish Server 中已经包含了 JSTL 的库，添加库的操作可以省略。所以选中刚才添加的两个.jar 文件并删除。

如果想要在 JSP 页面上使用 JSTL 标签，除了添加库文件以外，还需要在页面上使用 taglib 指令进行声明，语法为：

```
<%@taglib prefix="标签的前缀" uri="标签的 URI" %>
```

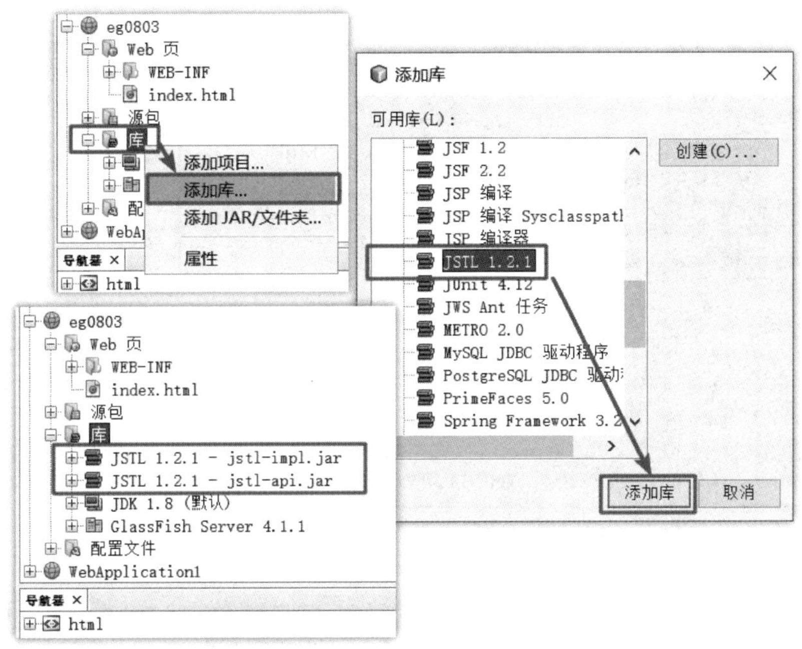

图 8.9　NetBeans 中添加 JSTL 库文件

例如，要使用核心库，则代码为：

```
<%@taglib prefix="c" uri="http://java.sun.com/jsp/jstl/core" %>
```

前缀及 URI 取值见表 8.8。

表 8.8　使用 JSTL 时前缀及 URI 取值

标签库	prefix	URI
core	c	http://java.sun.com/jsp/jstl/core
XML	x	http://java.sun.com/jsp/jstl/xml
I18N	fmt	http://java.sun.com/jsp/jstl/fmt
SQL	sql	http://java.sun.com/jsp/jstl/sql
Functions	fn	http://java.sun.com/jsp/jstl/functions

注意：对于 JSTL 1.0 版本，URI 中没有 jsp 这一级，形如 http://java.sun.com/jstl/core 等。

在 NetBeans IDE 中填写 uri 时，可以借助语法提示的快捷方式(Ctrl+\)来填充。在填写 uri 属性值时按快捷键，会给出语法提示，选择自己需要的就可以了，如图 8.10 所示。

在 eg0803 中分别新建下列 JSP 网页文件并在每个页面上都写好 taglib 声明，为下一步的学习做准备。

下一步的目标：学习 JSTL 核心标签库的使用。

工程名：eg0803。

用到的文件如表 8.9 所示。

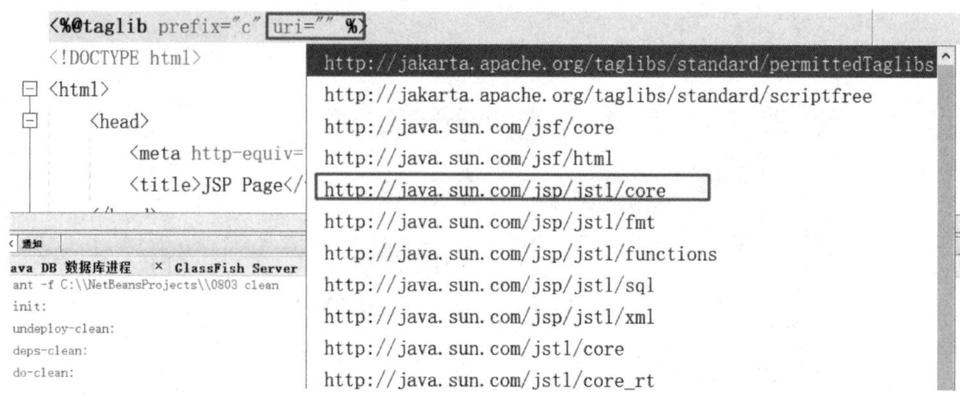

图 8.10 通过语法提示快捷键填写 uri

表 8.9 eg0803 用到的文件及文件说明

文件名	说明
forEach.jsp	手动创建的 JSP 文件,学习< forEach ></forEach >标签的使用
forTokens.jsp	手动创建的 JSP 文件,学习< forTokens ></forTokens >标签的使用
if.jsp	手动创建的 JSP 文件,学习< if ></if >标签的使用
choose.jsp	手动创建的 JSP 文件,学习< choose ></choose >标签的使用
chooseAndForEach.jsp	手动创建的 JSP 文件,学习将< choose >标签和< forEach ></forEach >标签结合在一起使用

3. JSTL 核心标签库(< c:forEach ></c:forEach >)

JSTL 核心标签库中控制循环的标签有< c:forEach ></c:forEach >和< c:forTokens ></c:forTokens >两组,下面先学习< c:forEach ></c:forEach >。

有关< c:forEach ></c:forEach >的使用,相关代码都放在 forEach.jsp 中。< c:forEach ></c:forEach >的用法有两种,一种与传统的 for 循环类似,需要知道初始条件、结束条件和步长。例如:想要输出 10 个"Hello!",Java 代码如下:

```
for ( int i = 1 ; i <= 10 ; i++){
    out.print("Hello!");
}
```

等价的< c:forEach ></c:forEach >代码如下,其中 begin 属性对应起始值,end 属性对应结束值,step 对应步长。

```
<h3>1 使用 forEach(begin、end、step)输出 10 个 Hello!</h3>
    <c:forEach begin = "1" end = "10" step = "1" >
        Hello!
    </c:forEach>
```

有时候需要控制循环变量,例如:想要输出 1~10 之间的整数,Java 代码如下:

```
for ( int i = 1 ; i <= 10 ; i++){
    out.print(i);
}
```

等价的< c:forEach ></c:forEach >代码如下，其中 begin 属性对应起始值，end 属性对应结束值，step 对应步长，而 var 属性的值为循环的对象变量名。

```
<h3>2 使用 forEach(var、begin、end、step)输出 1～10 之间的整数</h3>
    <c:forEach var = "output" begin = "1" end = "10" step = "1" >
        ${output}
    </c:forEach>
```

步长也可以取其他值，例如：

```
<h3>3 使用 forEach(var、begin、end、step)输出 1～10 之间的偶数</h3>
    <c:forEach var = "output" begin = "2" end = "10" step = "2">
        ${output}
    </c:forEach>
```

除了以传统的方式进行操作（往往是在循环次数已知的情况下），< c:forEach ></c:forEach >还提供了与增强型的 for 循环（for-each）对应的形式，作用是从某个集合中依次取出元素。例如下面的 Java 代码，从一个 String 数组中提取元素并输出。

```
String[] fruits = {"香蕉", "苹果", "哈密瓜"};
for(String s:fruits){
    out.print(s);
}
```

等价的< c:forEach ></c:forEach >代码如下，本例中 Java 脚本<% %>中先初始化了一个字符串数组 fruits，然后以 myfruits 的 key 名将其存储在 request 域中。< c:forEach ></c:forEach >的属性 items 代表待操作的集合，var 属性代表了当次循环取出的对象。其中属性中使用了 EL 语句"${myfruits}"以取出字符串数组 fruits。

```
<%
    String[] fruits = {"香蕉", "苹果", "哈密瓜"};
    request.setAttribute("myfruits", fruits);
%>
<h3>4 使用 forEach(var、items)输出集合中的元素</h3>
    <c:forEach var = "output" items = "${myfruits}">
        ${output}
    </c:forEach>
```

除了以上应用，< c:forEach ></c:forEach >还提供了一个属性 varStatus，该属性为循环状态变量，例如有时候在页面输出多条数据时，第一列往往起到计数的作用，此时可以使用${status.count}的形式显示计数，代表是第几条数据。请看下面的例子。

```
<h3>5 使用 forEach(var、items、varStatus)输出集合中的元素</h3>
    <c:forEach var = "output" items = "${myfruits}" varStatus = "status">
    第${status.count}个元素 ${output} <br/>
    </c:forEach>
```

以上关于<c:forEach></c:forEach>的用法，代码集中体现在 forEach.jsp 页面，全部代码如下：

forEach.jsp

```
<%@taglib prefix = "c" uri = "http://java.sun.com/jsp/jstl/core" %>
<%@page contentType = "text/html" pageEncoding = "UTF-8" %>
<!DOCTYPE html>
<html>
    <head>
        <meta http-equiv = "Content-Type" content = "text/html; charset = UTF-8">
        <title>使用 forEach </title>
    </head>
    <body>
      <h3>1 使用 forEach(begin、end、step)输出 10 个 Hello!</h3>
      <c:forEach begin = "1" end = "10" step = "1" >
          Hello!
      </c:forEach>
      <h3>2 使用 forEach(var、begin、end、step)输出 1～10 之间的整数</h3>
      <c:forEach var = "output" begin = "1" end = "10" step = "1" >
          ${output}
      </c:forEach>
      <h3>3 使用 forEach(var、begin、end、step)输出 1～10 之间的偶数</h3>
      <c:forEach var = "output" begin = "2" end = "10" step = "2">
          ${output}
      </c:forEach>
      <%
          String[] fruits = {"香蕉", "苹果", "哈密瓜"};
          request.setAttribute("myfruits", fruits);
      %>
      <h3>4 使用 forEach(var、items)输出集合中的元素</h3>
      <c:forEach var = "output" items = "${myfruits}">
          ${output}
      </c:forEach>
      <h3>5 使用 forEach(var、items、varStatus)输出集合中的元素</h3>
      <c:forEach var = "output" items = "${myfruits}" varStatus = "status">
          第${status.count}个元素 ${output} <br/>
      </c:forEach>
  </body>
</html>
```

表 8.10 总结了<c:forEach></c:forEach>标签的属性及作用。所有属性均为可选项，

而不是必须有。

表 8.10　< c:forEach ></c:forEach >标签的属性及作用

属性	描　　述
var	代表循环对象的变量名，若存在 items 属性，则表示循环集合中对象的变量名
items	进行循环的集合
varStatus	显示循环状态的变量
begin	开始条件
end	结束条件
step	循环的步长，默认为 1

4. JSTL 核心标签库(< c:forTokens ></c:forTokens >)

< c:forTokens ></c:forTokens >类似于 java. util. StringTokenizer 类，它提供了一种便捷的操纵字符串的方法。例如：有一个字符串为“吃饭，睡觉，打豆豆”中间的分隔符为“，”，如果想要取出“吃饭”“睡觉”“打豆豆”，那么可以采用< c:forTokens ></c:forTokens >。下面先看一个例子。

```
<h3>无意义的分隔(items、delims)</h3>
<c:forTokens items="吃饭,睡觉,打豆豆" delims=",">
Hello!
</c:forTokens>
```

这段代码将会输出 3 个“Hello!”。因为以逗号“，”分隔开的元素一共有 3 个，所以循环体执行 3 次。其中属性 items 代表要循环的对象，delims 代表分隔符。一般来说，需要输出被分隔的字符，所以< c:forTokens ></c:forTokens >提供了一个属性 var，用法与< c:forEach></c:forEach >标签中的 var 属性一样，代表循环对象的变量名，请看下面的代码。

```
<h3>输出分隔后元素(items、delims、var)</h3>
<c:forTokens items="吃饭,睡觉,打豆豆" delims="," var="s">
 ${s}
</c:forTokens>
```

这段代码将会输出“吃饭，睡觉，打豆豆”。如果想要输出分隔字符串中的一部分字符，那么可以结合使用 begin、end、step 属性，用法可参照< c:forEach ></c:forEach >。值得注意的是 begin 属性刻意设置为 1，代码运行将输出“睡觉，打豆豆”而不是“吃饭，睡觉”，说明下标是从 0 开始的。

```
<h3>输出分隔后元素中的某些元素<br/>(items、delims、var、begin、end、step)</h3>
<c:forTokens items="吃饭,睡觉,打豆豆" delims="," var="s" begin="1" end="2" step="1">
 ${s}
</c:forTokens>
```

完整的 forTokens.jsp 代码如下：

forTokens.jsp

```
<%@taglib prefix="c" uri="http://java.sun.com/jsp/jstl/core" %>
<%@page contentType="text/html" pageEncoding="UTF-8"%>
<!DOCTYPE html>
<html>
    <head>
        <meta http-equiv="Content-Type" content="text/html; charset=UTF-8">
        <title>JSP Page</title>
    </head>
    <body>
        <h3>1 无意义的分隔(items、delims)</h3>
        <c:forTokens items="吃饭,睡觉,打豆豆" delims=",">
            Hello!
        </c:forTokens>
        <h3>2 输出分隔后元素(items、delims、var)</h3>
        <c:forTokens items="吃饭,睡觉,打豆豆" delims="," var="s">
            ${s}
        </c:forTokens>
<h3>3 输出分隔后元素中的某些元素<br/>(items、delims、var、begin、end、step)</h3>
<c:forTokens items="吃饭,睡觉,打豆豆" delims="," var="s" begin="1" end="2" step="1"
>
            ${s}
</c:forTokens>
    </body>
</html>
```

5. JSTL 核心标签库(<c:if></c:if>)

JSTL 核心标签库中控制程序流程的标签有<c:if></c:if>、<c:choose></c:choose>两组，下面先学习<c:if></c:if>。

<c:if></c:if>标签提供了条件判断功能，判断条件取决于属性 test，此外<c:if></c:if>有两种形式，一种有标签体，一种没有，先看一下有标签体的情况。

```
<h3>1 最简单的条件判断及输出(test)</h3>
    <c:if test="${1<2}">
        条件为真
    </c:if>
```

这段代码输出字符串“条件为真”，因为 EL 表达式 ${1<2}的结果为 true，所以就输出了标签体里面的内容。

<c:if></c:if>第二种形式是没有标签体的，这时候必须有 var 属性，在标签体执行后，可以保存条件结果，默认是保存在当前页(page)中。

```
<h3>2 将判断结果赋值给变量并将该变量存储在默认作用域(test、var)</h3>
    <c:if test="${1<2}" var="output"/>
    ${pageScope.output}
```

除了将条件的结果保存在当前页(page)中,还可保存在 request、session、application 中,此时需要通过 scope 属性指明作用域。

```
<h3>3 将判断结果赋值给变量并将该变量存储在特定的作用域(test、var、scope)</h3>
    <c:if test="${2<2}" var="output" scope="request"/>
    ${requestScope.output}
```

完整的示例代码在 if.jsp 中,值得注意的是后两段代码片段,分别在 pageScope 和 requestScope 里存储了同名的(名字都为 output)变量,这里刻意将判断条件做了修改,以作对比。可以通过这个例子加深对 EL 的使用,例如:假如输出部分写成 ${output},那么输出结果会是什么呢?

if.jsp

```
<%@taglib prefix="c" uri="http://java.sun.com/jsp/jstl/core" %>
<%@page contentType="text/html" pageEncoding="UTF-8" %>
<!DOCTYPE html>
<html>
<head>
    <meta http-equiv="Content-Type" content="text/html; charset=UTF-8">
    <title>使用 if </title>
</head>
<body>
    <h3>1 最简单的条件判断及输出(test)</h3>
        <c:if test="${1<2}">
            条件为真
        </c:if>
    <h3>2 将判断结果赋值给变量并将该变量存储在默认的作用域(test、var)</h3>
        <c:if test="${1<2}" var="output"/>
        ${pageScope.output}
    <h3>3 将判断结果赋值给变量并将该变量存储在特定的作用域(test、var)</h3>
        <c:if test="${2<2}" var="output" scope="request"/>
        ${requestScope.output}
</body>
</html>
```

6. JSTL 核心标签库(<c:choose></c:choose>)

JSTL 中没有提供形如 if-else 这种形式的标签组,想达到类似的效果,可以通过<c:choose></c:choose>标签。<c:choose></c:choose>标签实际上是一组标签,包括了<c:choose></c:choose>、<c:when></c:when>、<c:otherwise></c:otherwise>三组标签,作用效果和开关语句(switch-case)相近。代码如下:

```
<%
    request.setAttribute("score",95);
%>
    <c:choose>
    <c:when test = "${score>= 90}">优</c:when>
    <c:when test = "${score>= 80}">良</c:when>
    <c:when test = "${score>= 70}">中</c:when>
    <c:when test = "${score>= 60}">及格</c:when>
    <c:otherwise>不及格</c:otherwise>
</c:choose>
```

这段代码首先在请求域中存储了一个名为 score，value 为 95 的 key 变量。而后将该变量与不同的分值作比较，将百分制转换成 5 分制，运行结果为"优"。也可以尝试将 95 改成其他数值，进行测试。

代码中< c:choose > </c:choose >作为最外层的父标签存在，没有属性。< c:when > </c:when>有一个属性 test，该属性用来设置判断的条件，< c:otherwise > </c:otherwise >表示当之前的条件均不满足的时候，默认执行的动作。完整的 choose. jsp 代码如下：

choose. jsp

```
<%@taglib prefix = "c" uri = "http://java.sun.com/jsp/jstl/core" %>
<%@page contentType = "text/html" pageEncoding = "UTF-8" %>
<!DOCTYPE html>
<html>
    <head>
        <meta http-equiv = "Content-Type" content = "text/html; charset = UTF-8">
        <title>使用 choose </title>
    </head>
    <body>
        <%
            request.setAttribute("score", 95);
        %>
        <c:choose>
            <c:when test = "${score>= 90}">优</c:when>
            <c:when test = "${score>= 80}">良</c:when>
            <c:when test = "${score>= 70}">中</c:when>
            <c:when test = "${score>= 60}">及格</c:when>
            <c:otherwise>不及格</c:otherwise>
        </c:choose>
    </body>
</html>
```

7. JSTL 核心标签库(< c:forEach >与< c:choose >结合)

下面的代码判断数组是否为空，如果为空则输出"没有满足条件的数据!"。否则，遍历输出数组中的各个元素。既要做判断又要做循环，所以需要将< c:forEach >与< c:choose >结合使用。

```
<h3>1 判断并输出数组中的元素</h3>
<%
    String[] fruits = {"香蕉", "苹果", "哈密瓜"};
    request.setAttribute("myfruits", fruits);
%>
<c:choose>
    <c:when test="${empty myfruits}">
        没有满足条件的数据!
    </c:when>
    <c:otherwise>
        <c:forEach var="fa" items="${myfruits}">
            ${fa}
        </c:forEach>
    </c:otherwise>
</c:choose>
```

下面的代码判断 List 是否为空，如果为空则输出"没有满足条件的数据!"。否则，遍历输出 List 中的各个元素。

```
<h3>2 判断并输出 List 中的元素</h3>
<%
    List<String> fruitList = new ArrayList();
    fruitList.add("芒果");
    fruitList.add("草莓");
    fruitList.add("水蜜桃");
    request.setAttribute("flist", fruitList);
%>
<c:choose>
    <c:when test="${empty flist}">
        没有满足条件的数据!
    </c:when>
    <c:otherwise>
        <c:forEach var="fb" items="${flist}">
            ${fb}
        </c:forEach>
    </c:otherwise>
</c:choose>
```

完整的 chooseAndForEach.jsp 文件代码如下，运行结果略。

chooseAndForEach.jsp

```
<%@page import="java.util.List" %>
<%@page import="java.util.ArrayList" %>
<%@taglib prefix="c" uri="http://java.sun.com/jsp/jstl/core" %>
<%@page contentType="text/html" pageEncoding="UTF-8" %>
<!DOCTYPE html>
<html>
```

```
<head>
    <meta http-equiv="Content-Type" content="text/html; charset=UTF-8">
    <title>使用 choose 和 forEach </title>
</head>
<body>
<h3>1 判断并输出数组中的元素</h3>
<%
    String[] fruits = {"香蕉", "苹果", "哈密瓜"};
    request.setAttribute("myfruits", fruits);
%>
<c:choose>
    <c:when test="${empty myfruits}">
        没有满足条件的数据!
    </c:when>
    <c:otherwise>
        <c:forEach var="fa" items="${myfruits}">
            ${fa}
        </c:forEach>
    </c:otherwise>
</c:choose>
<h3>2 判断并输出 List 中的元素</h3>
<%
    List<String> fruitList = new ArrayList();
    fruitList.add("芒果");
    fruitList.add("草莓");
    fruitList.add("水蜜桃");
    request.setAttribute("flist", fruitList);
%>
<c:choose>
    <c:when test="${empty flist}">
        没有满足条件的数据!
    </c:when>
    <c:otherwise>
        <c:forEach var="fb" items="${flist}">
            ${fb}
        </c:forEach>
    </c:otherwise>
</c:choose>
</body>
</html>
```

8.3 使用 JSTL、EL 改写案例 2

已经对 JSTL、EL 的最基本用法有了一定的了解，下面使用 JSTL 和 EL 对 eg0704 的案例进行改写。

复制 eg0704，在弹出的窗口中将项目名称改为 eg0804。这样就拥有了和 eg0704 完全相同的代码。改写 showCustomer.jsp 代码如下：

showCustomer.jsp

```
<%@taglib prefix="c" uri="http://java.sun.com/jsp/jstl/core" %>
<%@page contentType="text/html" pageEncoding="UTF-8"%>
<!DOCTYPE html>
<html>
    <head>
        <meta http-equiv="Content-Type" content="text/html; charset=UTF-8">
        <title>显示 customer 信息</title>
    </head>
    <body>
        <c:forEach var="c" items="${list}">
            customer_id : ${c.customer_id}
            name : ${c.name}
            <hr/>
        </c:forEach>
    </body>
</html>
```

可以看到 JSP 页面中已经完全没有 Java 脚本片段了。

8.4 使用 JSTL、EL 进一步改进案例 1 和案例 2

1. 再次改写案例 1

在 eg0802 中，使用 EL 改写了案例 1，当时只改写了 showCustomer.jsp 的代码，其他逻辑不变，下面尝试一下另一种处理方式。ConditionalQueryServlet 的代码片段如下：

```
if (rs.next()) {
    Customer c = new Customer();
    c.setCustomer_id(rs.getInt(1));
    c.setName(rs.getString(2));
    //以 key 名 c,存储了引用名为 c 的 Customer 对象
    request.setAttribute("c", c);
    request.getRequestDispatcher("showCustomer.jsp").forward(request, response);
} else {
    out.print("没有满足条件的 customer");
}
```

在 Servlet 中的输出语句是 out.print("没有满足条件的 customer")；而输出这个操作应该由 JSP 来承担才对。按照这个思路改变一下业务逻辑，即使没有满足条件的 customer，也将请求转发至 showCustomer.jsp 页面，相关提示信息在 showCustomer.jsp 页面输出。

复制 eg0802，将项目名称改为 eg0805，修改 ConditionalQueryServlet.java 后的代码

如下：

ConditionalQueryServlet.java 代码片段

```
try (PrintWriter out = response.getWriter()) {
    Class.forName(driver);
    conn = DriverManager.getConnection(url, user, password);
    prst = conn.prepareStatement(sql);
    prst.setInt(1, Integer.parseInt(customer_id));
    rs = prst.executeQuery();
    if (rs.next()) {
        Customer c = new Customer();
        c.setCustomer_id(rs.getInt(1));
        c.setName(rs.getString(2));
        //以 key 名 c,存储了引用名为 c 的 Customer 对象
        request.setAttribute("c", c);
    }
    request.getRequestDispatcher("showCustomer.jsp").forward(request, response);
} catch (ClassNotFoundException | SQLException ex) {
    Logger.getLogger(ConditionalQueryServlet.class.getName()).log(Level.SEVERE, null,
ex);
}
```

修改 showCustomer.jsp 后的代码如下：

```
<%@taglib prefix="c" uri="http://java.sun.com/jsp/jstl/core" %>
<%@page contentType="text/html" pageEncoding="UTF-8"%>
<!DOCTYPE html>
<html>
    <head>
        <meta http-equiv="Content-Type" content="text/html; charset=UTF-8">
        <title>显示 customer 信息</title>
    </head>
    <body>
        <c:choose>
            <c:when test="${empty c}">
                没有满足条件的 customer
            </c:when>
            <c:otherwise>
                customer_id: ${c.customer_id}<hr/>
                name: ${c.name}
            </c:otherwise>
        </c:choose>
    </body>
</html>
```

可见修改之后的代码，Servlet 中只有业务逻辑，没有输出任何响应。而 JSP 页面中只有输出，没有业务逻辑，也没有任何脚本，这使层次更为清晰。

2. 再次改写案例 2

用相同的思路再次改写案例 2。

复制 eg0804,将项目名称改为 eg0806,修改 ConditionalQueryServlet. java 代码如下:

ConditionalQueryServlet. java 代码片段

```
try (PrintWriter out = response.getWriter()) {
    Class.forName(driver);
    conn = DriverManager.getConnection(url, user, password);
    prst = conn.prepareStatement(sql);
    rs = prst.executeQuery();
    List<Customer> list = new ArrayList<>();
    //只要结果集不为空,就封装对象并添加至 list
    while (rs.next()) {
        Customer c = new Customer();
        c.setCustomer_id(rs.getInt(1));
        c.setName(rs.getString(2));
        list.add(c);
    }
    //以 key 名 list,存储了引用名为 list 的 List<Customer>对象
    request.setAttribute("list", list);
    request.getRequestDispatcher("showCustomer.jsp").forward(request, response);
} catch (ClassNotFoundException | SQLException ex) {
Logger.getLogger(ConditionalQueryServlet.class.getName()).log(Level.SEVERE, null, ex);
}
```

修改 showCustomer. jsp 代码如下:

```
<%@taglib prefix="c" uri="http://java.sun.com/jsp/jstl/core" %>
<%@page contentType="text/html" pageEncoding="UTF-8"%>
<!DOCTYPE html>
<html>
    <head>
        <meta http-equiv="Content-Type" content="text/html; charset=UTF-8">
        <title>显示 customer 信息</title>
    </head>
    <body>
        <c:choose>
            <c:when test="${empty list}">
                没有满足条件的 customer
            </c:when>
            <c:otherwise>
                <c:forEach var="c" items="${list}">
                    customer_id : ${c.customer_id}
                    name : ${c.name}
                    <hr/>
```

```
                </c:forEach>
            </c:otherwise>
        </c:choose>
    </body>
</html>
```

8.5 本章回顾

本章主要介绍了 EL 和 JSTL 的基础知识。EL 简化了页面输出，而 JSTL 则封装了常见的 JSP 应用程序的核心功能，提供了一套单一的、标准的标签组。本章先介绍了 EL 的用法，然后介绍了 JSTL 核心库中常用的几个标签。最后使用 JSTL 和 EL 改写了第 8 章的例子，可以看到，JSTL 和 EL 结合起来，能够大量减少 JSP 页面的 Java 脚本，提高了开发效率，增强了系统的维护性。本章小结如图 8.11 所示。

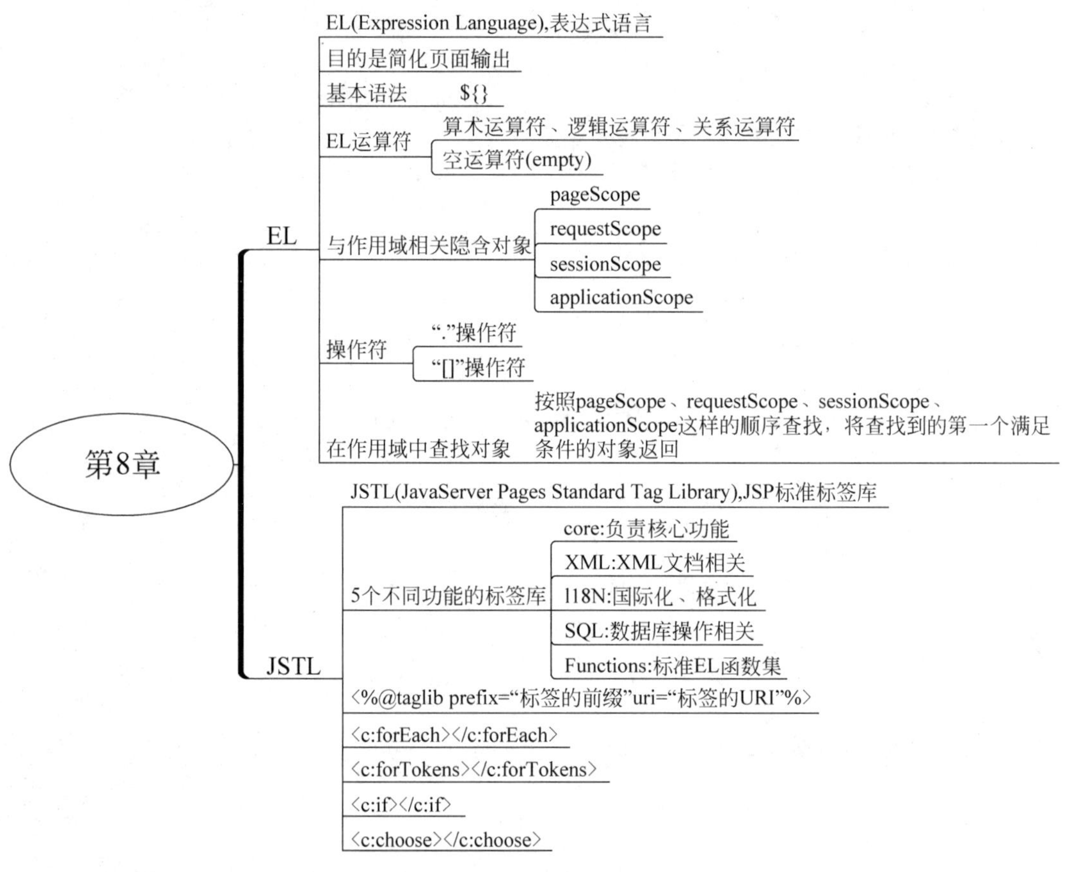

图 8.11 第 8 章内容结构图

8.6　课后习题

1. 名词解释：EL、JSTL。
2. EL 中与作用域相关的隐含对象有几个？分别是什么？
3. 在网页中使用 JSTL，需要提前做哪些设置？
4. 如果想在页面中使用 JSTL 核心库，taglib 应如何设置？请结合代码说明。

使用过滤器

学习目标：

通过本章的学习，你应该：

- 了解过滤器的基本原理
- 掌握使用过滤器解决乱码问题

9.1 过滤器概述

在开始过滤器的学习之前，先看两个生活中的场景，如图 9.1 和图 9.2 所示。老板和秘书的对话。

图 9.1　生活场景 1——拦截请求

因为老板给他的秘书设定的规则是："追债的就说我不在，还钱的就赶紧请进来"。所以当有人找老板时，秘书就按照这个规则来决定是"拦截请求"（见图 9.1），还是"放行请求"（见图 9.2）。图里面的秘书充当的就是一个过滤器的角色。Java EE 中的过滤器和这位秘书类似，它也可以拦截对服务器资源的请求，按照预定的规则来决定是否允许该请求通过。当然，对于过滤器而言，除了可以拦截请求，还可以拦截响应，在响应到达客户端之前做某种处理。

类比一下熟悉的校园生活，过滤器更像学校里的门卫。想象一下吧，有一个尽职尽责的门卫，所有想要进门的人都要接受检查，满足条件才允许进来；所有想出门的人，也要接受检查，满足条件才允许出去。而且，这样的门卫可以有多个，例如：学校大门的门卫、图书馆

图 9.2 生活场景 2——放行请求

的门卫、教学楼的门卫、宿舍楼的门卫，虽然每个门卫只管自己的这一亩三分地，但是他们组合在一起就控制了同学们的通行。例如：对于男同学而言，你可以通过学校大门去教学楼或去图书馆，但是你要想去女生宿舍，恐怕就会有问题，因为大多数学校的女生宿舍都会规定“男生止步”，你的通行请求会被禁止，只能打道回府。

Java EE 中的过滤器也是按照规则执行，一旦满足条件就自动触发，它主要完成两方面的任务：

(1) 在请求被送到目标之前检查请求对象，根据需要修改请求头和请求内容。

(2) 在响应被送到目标之前检查响应对象，根据需要修改响应头和响应内容。

每个过滤器都是独立的，它们可以经过配置来协同工作，构成一个过滤器链，如图 9.3 所示。在这个过滤器链条上，容器会根据过滤器配置情况来决定过滤器调用的次序，如果到了链尾，则会把控制权交给待访问的资源。等到该资源处理完毕，需要给客户端响应时，再逆序通过之前的过滤器链。

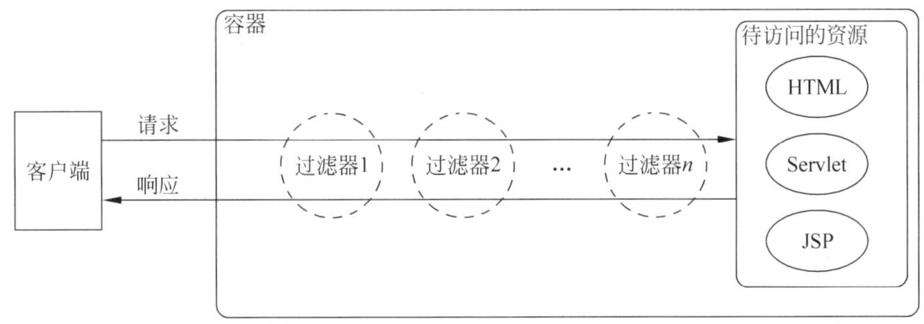

图 9.3 过滤器的基本原理图

更为巧妙的是，可以对过滤器进行自由组合，例如：假设有 10 个过滤器，过滤器 1、过滤器 2 到过滤器 10，还有资源 1、资源 2 和资源 3 三个资源，那么根据触发的条件不同，很可能出现这个状况，访问资源 1 使用了过滤器 1、3、5，访问资源 2 使用了过滤器 2、4、6，访问资源 3 使用了过滤器 1～10。只要设置好触发条件，这些过滤器就可以协同工作。只要编写好该过滤器触发后要做的动作，其他的就不用管了，整个过滤器链就会在容器的作用下自动运行了。

那么触发条件应该如何设置呢？又该在哪里编写过滤器设置被触发后该做的动作呢？下面举例说明。

9.2 过滤器的实现及部署

过滤器和 Servlet、JSP 一样，都是服务器端的一种组件，在容器中运行。从某些地方看，它和 Servlet 类似。表 9.1 展示了两者相似的地方。

表 9.1 Servlet 与 Filter 的比较

	Servlet	Filter
继承/实现接口	继承自 HttpServlet 类	实现了 Filter 接口
生命周期	初始化(调用 init 方法)	初始化(调用 init 方法)
	对外提供服务 (调用 Service 方法，由 Service 根据请求类型调用相应的 doXxx 方法，往往调用 doGet/doPost)	对外提供服务 (调用 doFilter 方法)
	销毁对象(调用 destory 方法)	销毁对象(调用 destory 方法)
部署方式 1 (使用配置文件)	<servlet> <servlet-name>x</servlet-name> <servlet-class>x</servlet-class> </servlet> <servlet-mapping> <servlet-name>x</servlet-name> <url-pattern>/x</url-pattern> </servlet-mapping>	<filter> <filter-name>x</filter-name> <filter-class>x</filter-class> </filter> <filter-mapping> <filter-name>x</filter-name> <url-pattern>/x</url-pattern> </filter-mapping>
部署方式 2 (使用 Annotation)	@WebServlet(name = "NewServlet", urlPatterns = {"/NewServlet"})	@WebFilter(filterName = "EncodingFilter", urlPatterns = {"/*"})

过滤器必须实现 Filter 接口，在 NetBeans IDE 中编写过滤器就像编写 Servlet 一样简单。编写 Servlet 时，主要是覆盖 doGet/doPost 方法(在 NetBeans IDE 中是编写 processRequest 方法)；编写过滤器时，主要是覆盖 doFilter 方法。

部署过滤器时也和部署 Servlet 类似，可以通过使用配置文件或使用 Annotation 两种方式进行。推荐使用 Annotation 的方式。部署过滤器的过程在 NetBeans IDE 中可以根据图形化的向导一步步进行，也很方便。

过滤器的部署实际上可以拆分成 3 个问题，解决了这 3 个问题，部署问题也就解决了，下面以一个简单的配置文件举例说明。

```
<filter>
    <filter-namer>x</filter-namer>
```

```
    <filter-class>x</filter-class>
</filter>
<filter-mapping>
    <filter-name>x</filter-name>
    <url-pattern>/x</url-pattern>
    <dispatcher>REQUEST</dispatcher>
    <dispatcher>FORWARD</dispatcher>
    <dispatcher>INCLUDE</dispatcher>
    <dispatcher>ERROR</dispatcher>
</filter-mapping>
```

过滤器是用来过滤对资源的访问的，那么第 1 个问题就是：如何去描述某个资源？答案很简单，通常都是< url-pattern >，任何一个资源都可以抽象成一个< url-pattern >。

第 2 个问题就是：如何去描述某个过滤器呢？答案也很简单，就是通过过滤器的逻辑名< filter-name >，以及该逻辑名对应的实际的类< filter-class >；通过过滤器的逻辑名可以将< url-pattern >和实际的过滤器类< filter-class >联系起来。

第 3 个问题就是对待资源的请求方式：对这个资源或者这些资源的请求时，是直接发送请求(REQUEST)呢？还是通过转发器转发(FORWARD)过来的请求呢？或是通过转发器转发(INCLUDE)过来的请求呢？还是通过声明式异常处理机制调用(ERROR)发送来的请求呢？请求的方式通过< dispatcher >来描述。默认是 REQUEST，意即没有特别说明< dispatcher >，都按 REQUEST 方式处理。

采用 Annotation 方式，也是描述了这 3 个方面的问题，可能的形式如下：

```
@WebFilter(
    filterName = "x",
    urlPatterns = {"/x"}
    dispatcherTypes = {
        DispatcherType.FORWARD,
        DispatcherType.ERROR,
        DispatcherType.REQUEST,
        DispatcherType.INCLUDE
    }
)
```

在本书的例子中，主要采用 Annotation 方式对过滤器进行部署。

9.3 在项目中使用一个过滤器

下面通过一个简单的示例学习过滤器的使用，过滤对某个特定资源的访问。

目标：学习使用 Filter，理解 Filter 的运行过程。

工程名：eg0901。

用到的文件如表 9.2 所示。

表 9.2 eg0901 用到的文件及文件说明

文件名	说明
index.jsp	创建工程后手动创建的文件,对其进行简单改写即可。在控制台输出提示信息
NewFilter.java	手动创建的 Filter,用来过滤对 index.jsp 的访问请求

打开 NetBeans IDE,新建一个名为 eg0901 的 Java Web 项目,具体过程略;删除 index.html 文件,新建名为 index.jsp 的文件,具体过程略;新建名为 NewFilter 的 Filter,过程如下:

打开项目 eg0901,在源包上右击,选择"新建"命令。第一次创建时,并没有过滤器选项,这时选择"其他"命令,如图 9.4 所示。

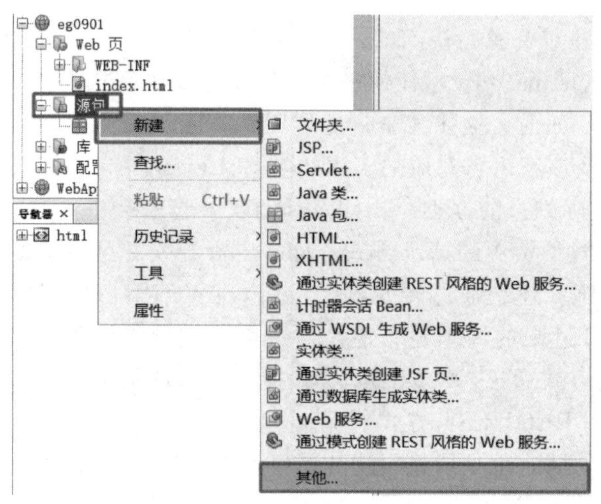

图 9.4 选择"其他"命令

在弹出的选择文件类型窗口中选择"类别"为 Web,在 Web 类别下的"文件类型"中选择"过滤器"类型,然后单击"下一步"按钮,如图 9.5 所示。

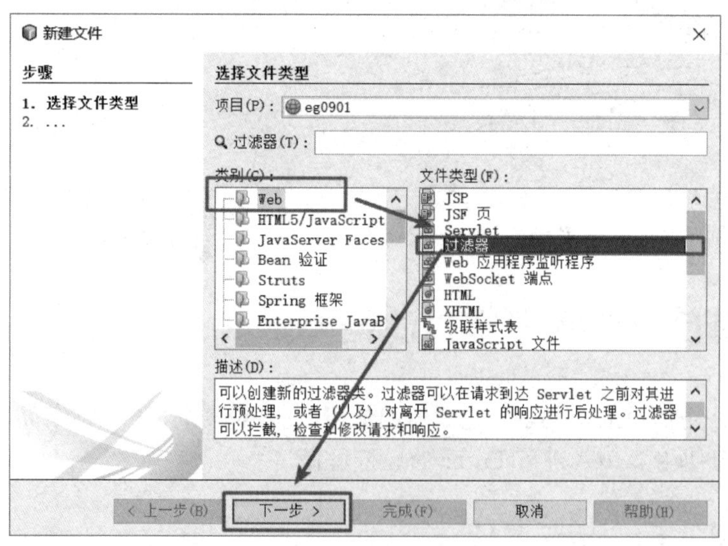

图 9.5 新建过滤器

在弹出的名称和位置窗口中设置类名为 NewFilter，包名为 cn. edu. djtu. util（因为过滤器也属于整个项目共用的工具类，所以放在 util 包中），设置完成后单击“下一步”按钮，如图 9.6 所示。

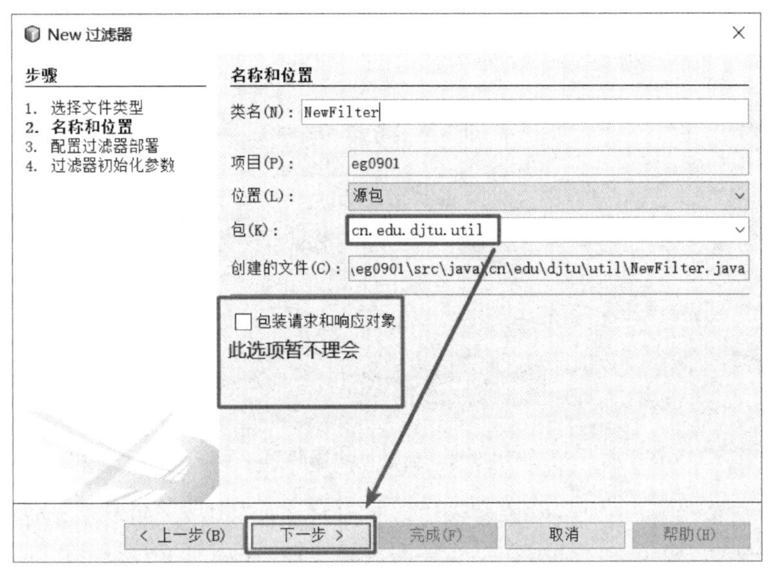

图 9.6 设置过滤器的包名

选择之后会弹出新的“配置过滤器部署”的窗口，在该窗口中可以看到“将信息添加到部署描述符(web. xml)”单选框。按照传统的部署方式，是要将该信息添加到部署描述符的，如果不勾选，则意味着采用注解方式部署，这里采用注解方式，所以不勾选该选项。

屏幕下方红色的感叹号提示要“至少输入一种 URL 模式”，因此单击“新建”按钮，设置 URL 模式，如图 9.7 所示。

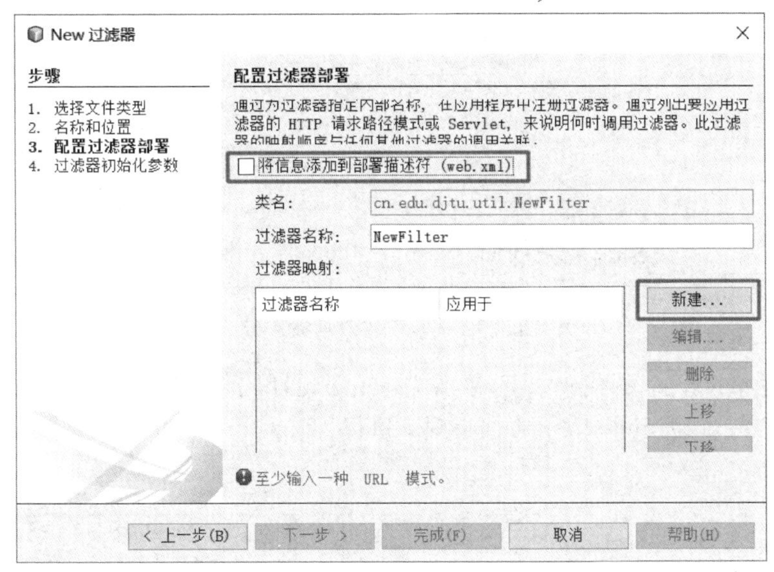

图 9.7 配置过滤器部署

在弹出的“过滤器映射”窗口中可以看到有两个地方可以设置，一个是URL，默认的形式是“/*”，其中*号为通配符，表示所有的请求。另一个地方是分发条件，可以选择4个中的某一个或者某几个，如果不选择，则默认相当于选中了REQUEST。如图9.8左图所示。

本次创建的过滤器是过滤针对index.jsp页面的请求，所以URL应该设置为“/index.jsp”，“分派条件”全选，则需要设置成图9.8右图形式。

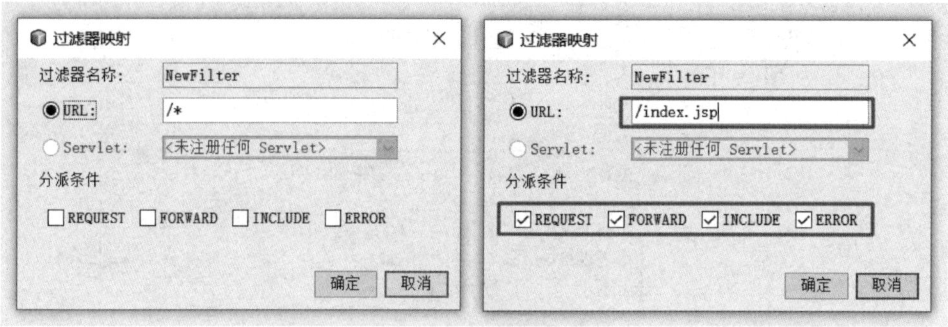

图9.8 配置过滤器映射

配置完成后单击“确定”按钮，则基本完成了对过滤器的设置，如图9.9所示。如果此时单击“下一步”按钮，则进入过滤器初始化参数设置界面，此处单击“完成”按钮。

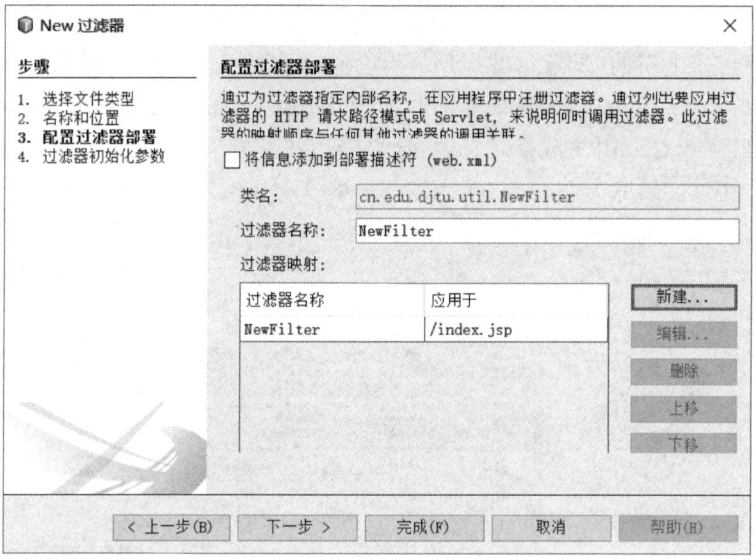

图9.9 完成配置过滤器部署

首先编写index.jsp的代码如下，添加了一条语句，用于在控制台输出，当在控制台看到index.jsp字样时，说明index.jsp页面得到了执行。

index.jsp

```
<%@page contentType="text/html" pageEncoding="UTF-8"%>
<!DOCTYPE html>
```

```
<html>
    <head>
        <meta http-equiv="Content-Type" content="text/html; charset=UTF-8">
        <title>JSP Page</title>
    </head>
    <body>
        <%
            System.out.println("index.jsp");
         %>
    </body>
</html>
```

接下来编写 NewFilter.java 的代码，前面说过，过滤器(Filter)的生命周期和 Servlet 类似分成了三个阶段，主要编写 doFilter 方法，那么下面就看一下这个方法。在 NetBeans IDE 中编写过滤器，看起来有一点乱，这是为了编程的灵活性，NetBeans IDE 自定义了一些方法，其实真正写代码并不复杂。在项目下面的 doFilter 导航中可以看到过滤器中有哪些方法和属性。在 doFilter 方法上面单击，可以快速进入到该方法的方法体。图 9.10 中 doFilter 方法有三个参数，类型分别为 ServletRequest、ServletResponse 和 FilterChain，代表了过滤器中需要的三个对象——请求对象、响应对象和过滤器链对象。

图 9.10 过滤器中的方法和属性

先看一下 doFilter 的代码片段：

```
public void doFilter(ServletRequest request, ServletResponse response,FilterChain chain)
    throws IOException, ServletException {
    //代码略
    doBeforeProcessing(request, response);
```

```
    //代码略
    try {
        chain.doFilter(request, response);
    } catch (Throwable t) {
        //代码略
    }
    doAfterProcessing(request, response);
    //代码略
    }
```

可以看到主要的语句其实就是 3 条，以 chain.doFilter(request,response);为界，chain.doFilter(request, response);之前是对请求进行过滤，chain.doFilter(request,response);之后是对响应进行过滤。而 chain.doFilter(request,response); 本身则是指明了要调用该过滤器链条上的下一个过滤器。NetBeans IDE 中分别定义了 doBeforeProcessing(request, response); 和 doAfterProcessing(request, response); 两个方法来完成任务，这两个方法并不是 Filter 接口中的方法。

为了更好地看到过滤器执行时的顺序，在 NewFilter 的 doFilter 方法中添加波浪线所示的代码。

NewFilter.java 的 doFilter 方法(片段)

```
public void doFilter(ServletRequest request, ServletResponse response,FilterChain chain)
    throws IOException, ServletException {
    //代码略
    doBeforeProcessing(request, response);
    //代码略
try {
    System.out.println("在 NewFilter 的 chain.doFilter(request, response); 语句前");
    chain.doFilter(request, response);
    System.out.println("在 NewFilter 的 chain.doFilter(request, response); 语句后");
} catch (Throwable t) {
        //代码略
    }
    doAfterProcessing(request, response);
    //代码略
    }
```

然后运行 index.jsp，按照前面的知识，预期的输出及相应顺序应该是先执行 chain.doFilter(request, response); 之前的语句，输出“在 NewFilter 的 chain.doFilter(request, response); 语句前”; 然后执行 chain.doFilter(request, response); 语句，调用链条上的下一个过滤器，如果已经到了过滤器的末端，则将控制权交给待访问的资源，此处只有一个过滤器，因此执行 index.jsp 的代码输出“index.jsp”; 最后再按逆序遍历各个过滤器，输出“在 NewFilter 的 chain.doFilter(request, response); 语句后”，执行顺序如图 9.11 所示。

观察控制台输出可以看到，与预期一致，如图 9.12 所示。

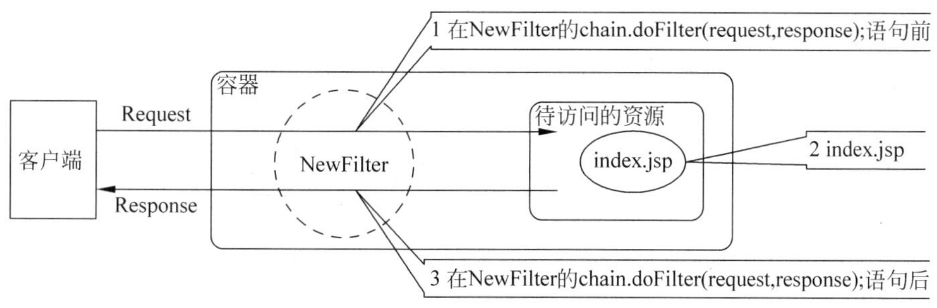

图 9.11 eg0901 预期的执行顺序

图 9.12 控制台输出

9.4 在项目中使用多个过滤器

在项目 eg0901 中学习了如何使用过滤器，但只涉及一个过滤器，那么当多个过滤器映射到同一个给定的资源时，过滤器又该如何执行呢？下面再通过一个例子来看一下。

通过复制项目 eg0901 的方式创建项目 eg0902。已知在项目 eg0901 中有一个过滤器 NewFilter。当运行 index.jsp 时，过滤器 NewFilter 满足条件，应该起作用。

新建一个过滤器 NewFilter1，配置 NewFilter1 的过滤器映射时，将 URL 设置为“*.jsp”，如图 9.13 所示。注意和前面 NewFilter1 配置有不同，不是“/*.jsp”。这意味着所有

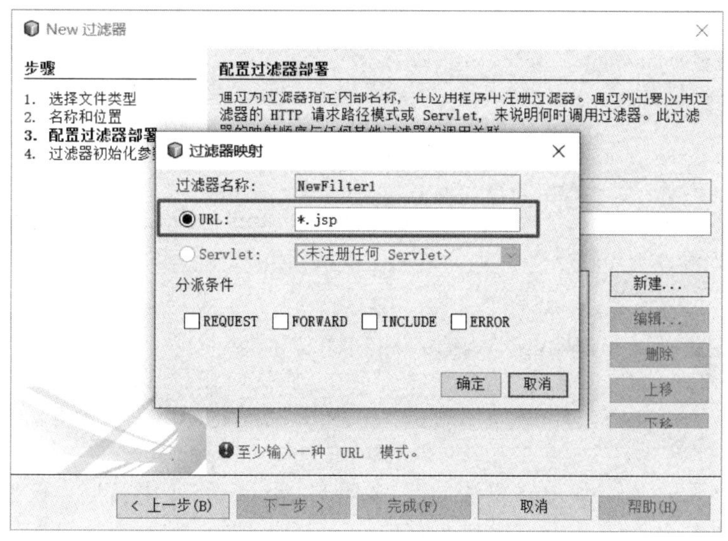

图 9.13 配置 NewFilter1 的过滤器映射

以“.jsp”结尾的 url-pattern 访问，都会经过该过滤器，例如：a.jsp、b.jsp、c.action.jsp。所以运行 index.jsp 时，过滤器 NewFilter1 满足条件，应该起作用。

新建一个过滤器 NewFilter2，配置 NewFilter2 的过滤器映射时，将 URL 设置为“/＊”，如图 9.14 所示。这意味着所有请求都会经过该过滤器，例如：a.action、NewServlet、c.jsp、d.html。所以当运行 index.jsp 时，过滤器 NewFilter2 满足条件，应该起作用。

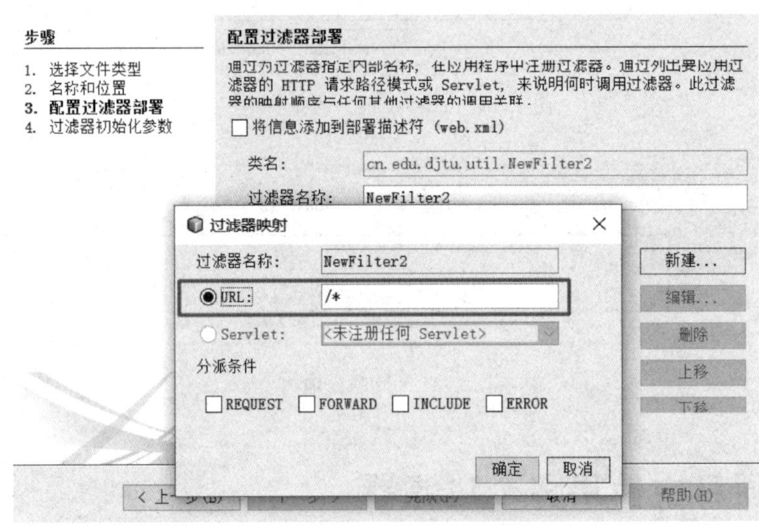

图 9.14 配置 NewFilter2 的过滤器映射

根据以上设定，index.jsp 运行时，NewFilter、NewFilter1 和 NewFilter2 都会起作用。

编写 NewFilter1 的 doFilter 方法如下，逻辑与 NewFilter 中类似。仅呈现代码片段。

NewFilter1.java 的 doFilter 方法(片段)

```
public void doFilter(ServletRequest request, ServletResponse response,FilterChain chain)
        throws IOException, ServletException {
    //代码略
    doBeforeProcessing(request, response);
    //代码略
try {
    System.out.println("在 NewFilter1 的 chain.doFilter(request, response); 语句前");
    chain.doFilter(request, response);
    System.out.println("在 NewFilter1 的 chain.doFilter(request, response); 语句后");
} catch (Throwable t) {
    //代码略
    }
    doAfterProcessing(request, response);
    //代码略
    }
```

编写 NewFilter2 的 doFilter 方法如下，逻辑与 NewFilter1 中类似。仅呈现代码片段。

NewFilter2.java 的 doFilter 方法(片段)

```
public void doFilter(ServletRequest request, ServletResponse response,FilterChain chain)
    throws IOException, ServletException {
    //代码略
    doBeforeProcessing(request, response);
    //代码略
try {
    System.out.println("在 NewFilter2 的 chain.doFilter(request, response); 语句前");
    chain.doFilter(request, response);
    System.out.println("在 NewFilter2 的 chain.doFilter(request, response); 语句后");
} catch (Throwable t) {
    //代码略
    }
    doAfterProcessing(request, response);
    //代码略
    }
```

运行 index.jsp，观察控制台输出，可以看到 3 个过滤器执行的顺序为 NewFilter1、NewFilter、NewFilter2，如图 9.15 所示。

```
信息:   WebModule[null] ServletContext.log():NewFilter1:doFilter()
信息:   WebModule[null] ServletContext.log():NewFilter1:DoBeforeProcessing
信息:   在NewFilter1的chain.doFilter(request, response);语句前
信息:   WebModule[null] ServletContext.log():NewFilter:doFilter()
信息:   WebModule[null] ServletContext.log():NewFilter:DoBeforeProcessing
信息:   在NewFilter的chain.doFilter(request, response);语句前
信息:   WebModule[null] ServletContext.log():NewFilter2:doFilter()
信息:   WebModule[null] ServletContext.log():NewFilter2:DoBeforeProcessing
信息:   在NewFilter2的chain.doFilter(request, response);语句前
信息:   index.jsp
信息:   在NewFilter2的chain.doFilter(request, response);语句后
信息:   WebModule[null] ServletContext.log():NewFilter2:DoAfterProcessing
信息:   在NewFilter的chain.doFilter(request, response);语句后
信息:   WebModule[null] ServletContext.log():NewFilter:DoAfterProcessing
信息:   在NewFilter1的chain.doFilter(request, response);语句后
信息:   WebModule[null] ServletContext.log():NewFilter1:DoAfterProcessing
```

图 9.15 执行多个过滤器

从运行结果可以看到过滤器运行时不需要主动调度，只要满足过滤器设置的过滤条件，它们就会自动执行。而且执行的过程与图 9.3 预测的相同。

此外，观察一下图 9.13，其实配置过滤器映射时还可以按照 Servlet 来过滤，只是考虑到即使是 Servlet，也是按照某个 url-pattern 来进行访问，因此多数情况下还是设置 URL 居多。

9.5 使用过滤器处理中文乱码

在 3.9 节中学习了一种中文乱码的解决方法，这种方法可以解决用户通过 post 方式提交中文信息时产生的乱码，语句很简单。

```
request.setCharacterEncoding("UTF-8");
```

但是比较麻烦的是，只要涉及类似场景，就要在代码中添加该条语句。下面采用过滤器来解决这个问题，避免该条语句在代码中的蔓延。开发时简单，维护时也比较容易。

目标：使用 Filter 解决中文乱码。

工程名：eg0903（通过复制项目 eg0304 的方式创建项目 eg0903）。

用到的文件如表 9.3 所示。

表 9.3 eg0903 用到的文件及文件说明

文件名	说明
index.html	创建工程时自动创建。为项目首页。页面上有一个表单。表单内有一个文本框，一组复选框
ShowFormServlet.java	手动创建的 Servlet，用来取得用户通过表单提交的信息，并以网页形式显示
EncodingFilter.java	手动创建的过滤器，用来设置请求的编码格式

打开 NetBeans IDE，复制项目 eg0304 并设置项目名为 eg0903，具体过程略；

将 ShowFormServlet.java 代码中的语句 request.setCharacterEncoding ("UTF-8"); 删除。

ShowFormServlet.java 代码片段（删除处理中文乱码的语句后）

```
protected void processRequest(HttpServletRequest request, HttpServletResponse response)
    throws ServletException, IOException {
        response.setContentType("text/html;charset=UTF-8");
        try (PrintWriter out = response.getWriter()) {
            //获得 msg 的值并输出
            String msg = request.getParameter("msg");
            out.print(msg);
            out.print("<hr />");
            //获得 fruits 的值,如果用户未作选择,提示"未做选择"
            //否则遍历输出
            String[] fruits = request.getParameterValues("fruits");
            if (fruits == null) {
                out.print("未做选择");
            } else {
                for (String fruit : fruits) {
                    out.print(fruit + "<br />");
                }
            }
    }
}
```

在 cn.edu.djtu.util 的包中新建名为 EncodingFilter.java 的 Filter。URL 设置为"/*"，其他选项取默认值，具体过程略。在创建的 EncodingFilter 的 doFilter 方法中添加处理中文乱码的语句："request.setCharacterEncoding("UTF-8");"，代码片段如下：

EncodingFilter.java 的 doFilter 方法(片段)

```
public void doFilter(ServletRequest request, ServletResponse response,FilterChain chain)
    throws IOException, ServletException {
    //代码略
    doBeforeProcessing(request, response);
    //代码略
    try {
        request.setCharacterEncoding("UTF-8");
        chain.doFilter(request, response);
    } catch (Throwable t) {
        //代码略
    }
    doAfterProcessing(request, response);
    //代码略
    }
```

这样一来,所有的请求,编码就都被设置成 UTF-8 了,只需要写一条就可以了。运行后可见乱码问题得到了解决。

9.6 本章回顾

本章先介绍了过滤器的概念,然后探讨了过滤器的基本用法,以及如何在 NetBeans 中使用过滤器,过滤器的执行顺序等问题,最后通过例子展示了如何通过过滤器处理中文乱码问题。通过本章的学习,可以看到,过滤器可以对进入系统的请求和离开系统的响应进行过滤,对请求进行预处理,对响应进行后处理,集中解决全局性的问题。同时,使用过滤器比较简单,一般只需要在 doFilter 中进行编程就可以了。本章小结如图 9.16 所示。

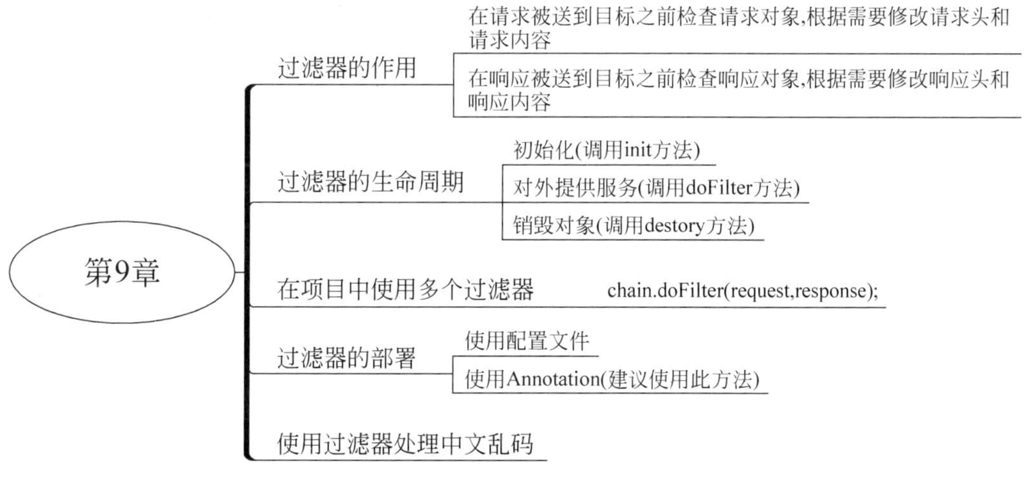

图 9.16 第 9 章内容结构图

9.7 课后习题

1. 过滤器的主要作用?
2. 过滤器与 Servlet 有什么区别与联系?
3. 编写过滤器时,一般在哪个方法中实现自己的业务逻辑?(写出方法名及参数)
4. 部署过滤器有几种方法?分别是什么?试结合代码举例说明。

第10章 DAO设计模式

学习目标：

通过本章的学习，你应该：

- 掌握 VO 设计模式
- 掌握 DAO 设计模式
- 掌握工厂设计模式
- 掌握 MVC 设计模式

10.1 DAO 设计模式案例需求分析

本章将通过一个功能简单的 MIS(管理信息系统)来学习 DAO 设计模式。在学习过程中，业务逻辑不变，但是系统实现方式将经历多次迭代。在实现过程中要注意比较不同实现方式的差异，从而对设计模式有个直观的感受。

系统目前业务逻辑比较简单。该系统的参与者为管理员，管理员使用“登录”和“添加用户”两个用例，如图 10.1 所示。

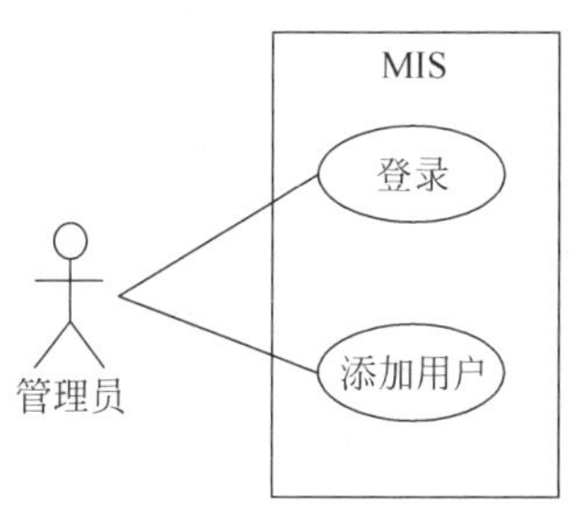

图 10.1　MIS 用例图

登录的活动图如图 10.2 所示。管理员进入 login.jsp 页面，login.jsp 页面输出请求域内的提示信息以及登录表单。管理员输入用户名和密码，并将表单提交至 LoginHandleServlet。LoginHandleServlet 获得管理员提交的用户名和密码后判断用户名和密码是否匹配，如果匹配则将用户名存入会话域，并将请求转发至 index.jsp，由 index.jsp 显示欢迎信息；如果用户名和密码不匹配，则将错误提示信息存入请求域，并将请求转发至 login.jsp 重新登录。

“添加用户”的活动图如图 10.3 所示。管理员登录成功后，进入 index.jsp 页面。index.jsp 页面显示首页信息。管理员单击“添加用户”的超级链接，跳转到 addLogin.jsp。addLogin.jsp 输出请求域中的提示信息，显示添加用户的表单，管理员输入用户信息并将请求提交至 AddLoginServlet。AddLoginServlet 判断该用户是否已存在，如果已存在，则将提示信息存入请求域，并将请求转发至 addLogin.jsp；如果该用户不存在，则添加该用户，并将请求转发至首页 index.jsp。

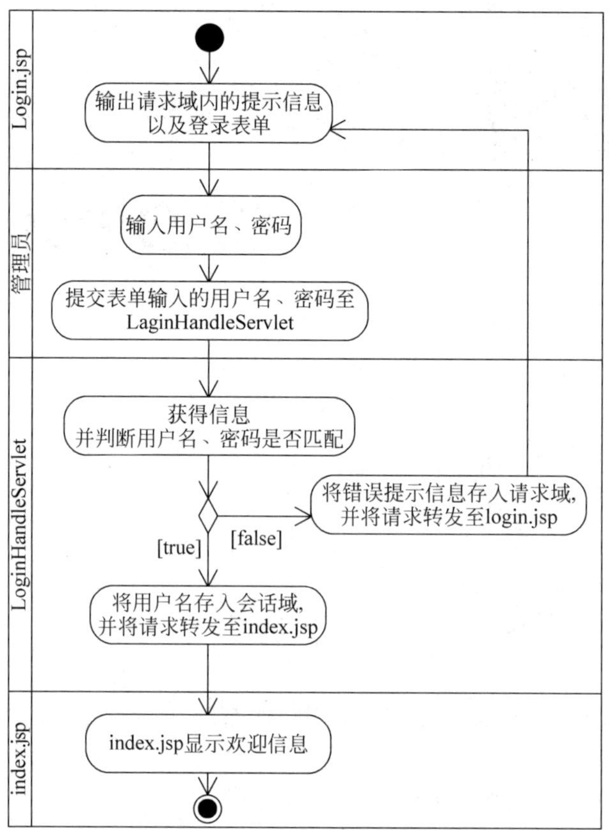

图 10.2 登录的活动图

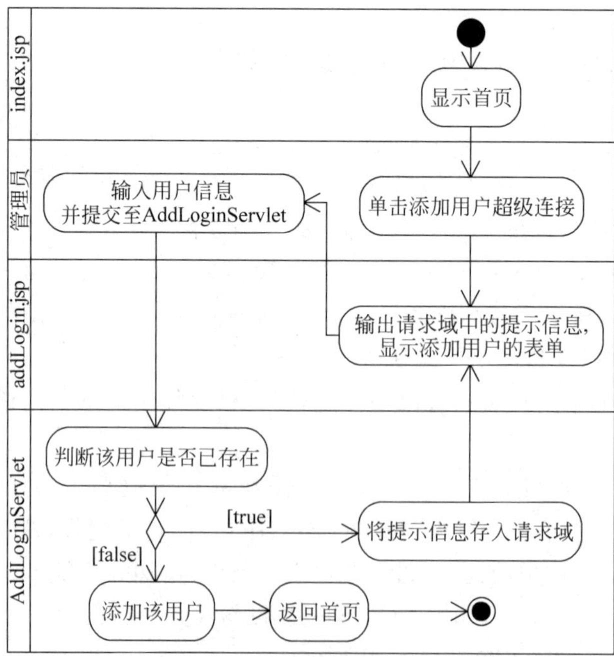

图 10.3 添加用户的活动图

10.2 数据库设计与实现

依然使用样例数据库 sample，在样例数据库中添加一张 login 表，用来存储登录的管理员信息，login 表的结构如表 10.1 所示。

表 10.1 login 表

字段名	数据类型	约束	备注
username	VARCHAR(15)	PRIMARY KEY	登录用户名
userpass	VARCHAR(8)	NULL	登录密码

建表语句如下：

login. sql

```
CREATE TABLE login(
    usernameVARCHAR(15) ,
    userpassVARCHAR (8) not null ,
    CONSTRAINT pk_userinfo PRIMARY KEY(username)
)
```

选择“服务”，打开 sample 数据库连接，展开 App，右击“表”，选择“执行命令”命令，如图 10.4 所示。

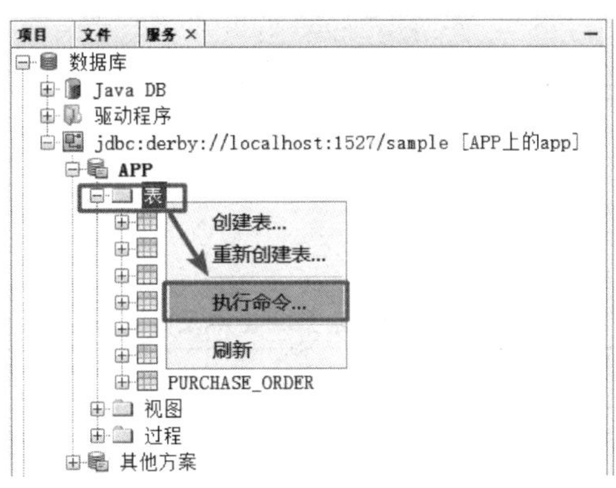

图 10.4 选择“执行命令”

在右侧弹出的窗口中编写建表语句 login. sql 并执行，如图 10.5 所示。

按照同样的方式执行命令，在 login 表中录入数据，初始化 login 表（为了简化逻辑，此处密码采用明文存放，实际工程应加密存储）。

```
INSERT INTO login VALUES('admin','admin');
```

至此，数据库设计与实现全部完成。右击“LOGIN”表，选择“查看数据”命令，可以看到 login 表中已有的数据，如图 10.6 所示。

```
SQL 3 [jdbc:derby://localhost:15…]
连接(O): jdbc:derby://localhost:1527/sample [APP上的app]
CREATE TABLE login(
    username VARCHAR(15) ,
    userpass VARCHAR (8) not null ,
    CONSTRAINT pk_userinfo PRIMARY KEY(username)
)
```

运行sql语句

图 10.5 执行创建 login 表的语句

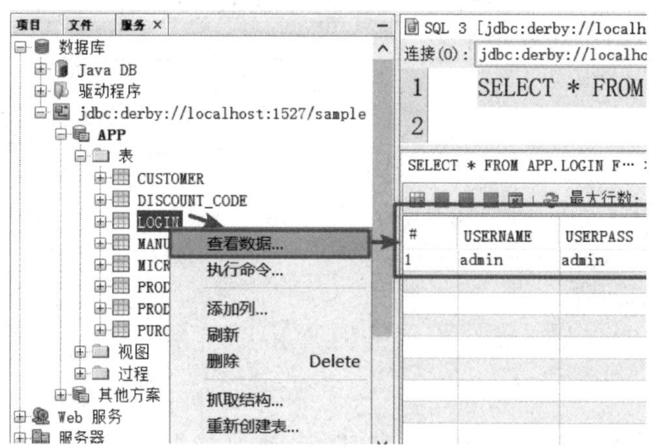

图 10.6 查看 login 表数据

10.3 MIS 第 1 版实现

工程名：eg1001。

用到的文件如表 10.2 所示。

表 10.2 eg1001 用到的文件及文件说明

文 件 名	说 明
login.jsp	用户登录页
index.jsp	首页
LoginHandleServlet.java	处理用户登录的 Servlet,位于 cn.edu.djtu 包中
addLogin.jsp	添加新用户的页面
AddLoginServlet.java	处理添加用户的 Servlet,位于 cn.edu.djtu 包中

打开 NetBeans IDE,新建一个名为 eg1001 的 Java Web 项目,具体过程略；删除 index.html 文件,新建各文件并编写代码如下：

login.jsp

```
<%@page contentType = "text/html" pageEncoding = "UTF-8" %>
<!DOCTYPE html>
```

```
<html>
    <head>
        <meta http-equiv="Content-Type" content="text/html; charset=UTF-8">
        <title>用户登录</title>
    </head>
    <body>
        ${error}
        <form action="LoginHandleServlet" method="post">
            用户名:<input type="text" name="username"/><br/>
            密码:<input type="password" name="userpass"/><br/>
            <input type="submit" value="登录"/>
        </form>
    </body>
</html>
```

addLogin.jsp

```
<%@page contentType="text/html" pageEncoding="UTF-8"%>
<!DOCTYPE html>
<html>
    <head>
        <meta http-equiv="Content-Type" content="text/html; charset=UTF-8">
        <title>添加用户</title>
    </head>
    <body>
        ${error}
        <form action="AddLoginServlet" method="post">
            用户名:<input type="text" name="username"/><br/>
            密码:<input type="password" name="userpass"/><br/>
            <input type="submit" value="添加用户"/>
        </form>
    </body>
</html>
```

index.jsp

```
<%@page contentType="text/html" pageEncoding="UTF-8"%>
<!DOCTYPE html>
<html>
    <head>
        <meta http-equiv="Content-Type" content="text/html; charset=UTF-8">
        <title>首页</title>
    </head>
    <body>
        <h1>Hello ${username}!</h1>
        <a href="addLogin.jsp">添加用户</a>
    </body>
</html>
```

LoginHandleServlet. java

```
package cn.edu.djtu;

import java.io.IOException;
import java.io.PrintWriter;
import java.sql.Connection;
import java.sql.DriverManager;
import java.sql.PreparedStatement;
import java.sql.ResultSet;
import java.sql.SQLException;
import java.util.logging.Level;
import java.util.logging.Logger;
import javax.servlet.ServletException;
import javax.servlet.annotation.WebServlet;
import javax.servlet.http.HttpServlet;
import javax.servlet.http.HttpServletRequest;
import javax.servlet.http.HttpServletResponse;
import javax.servlet.http.HttpSession;

@WebServlet(name = "LoginHandleServlet", urlPatterns = {"/LoginHandleServlet"})
public class LoginHandleServlet extends HttpServlet {

    /**
     * Processes requests for both HTTP <code>GET</code> and <code>POST</code>
     * methods.
     *
     * @param request servlet request
     * @param response servlet response
     * @throws ServletException if a servlet-specific error occurs
     * @throws IOException if an I/O error occurs
     */
    protected void processRequest(HttpServletRequest request, HttpServletResponse response)
            throws ServletException, IOException {
        response.setContentType("text/html;charset=UTF-8");
        String username = request.getParameter("username");
        String userpass = request.getParameter("userpass");

        //连接数据库用到的对象
        Connection conn = null;
        PreparedStatement prst = null;
        ResultSet rs = null;

        //连接数据库用到的参数信息
        String url = "jdbc:derby://localhost:1527/sample";
        String driver = "org.apache.derby.jdbc.ClientDriver";
        String user = "app";
        String password = "app";
```

```
        //查询数据库的 SQL 语句
        String sql = "SELECT username , userpass FROM login WHERE username = ?";
        try (PrintWriter out = response.getWriter()) {
            Class.forName(driver);
            conn = DriverManager.getConnection(url, user, password);
            prst = conn.prepareStatement(sql);
            prst.setString(1, username);
            rs = prst.executeQuery();
            if (rs.next() && rs.getString(2).equals(userpass)) {
                HttpSession session = request.getSession();
                session.setAttribute("username",username);
                request.getRequestDispatcher("index.jsp").forward(request, response);
            } else {
                request.setAttribute("error", "用户名密码错误");
                request.getRequestDispatcher("login.jsp").forward(request, response);
            }
        } catch (ClassNotFoundException | SQLException ex) {
        Logger.getLogger(LoginHandleServlet.class.getName()).log(Level.SEVERE, null, ex);
        }finally{
            //关闭结果集对象
            if (rs != null) {
                try {
                    rs.close();
                } catch (SQLException ex) {
Logger.getLogger(LoginHandleServlet.class.getName()).log(Level.SEVERE, null, ex);
                }
            }
            //关闭预编译语句对象
            if (prst != null) {
                try {
                    prst.close();
                } catch (SQLException ex) {
Logger.getLogger(LoginHandleServlet.class.getName()).log(Level.SEVERE, null, ex);
                }
            }
            //关闭连接对象
            if (conn != null) {
                try {
                    conn.close();
                } catch (SQLException ex) {
Logger.getLogger(LoginHandleServlet.class.getName()).log(Level.SEVERE, null, ex);
                }
            }
        }
    }

    // <editor-fold defaultstate="collapsed" desc="HttpServlet methods. Click on the +
sign on the left to edit the code.">
    /**
```

```
     * Handles the HTTP <code>GET</code> method.
     *
     * @param request servlet request
     * @param response servlet response
     * @throws ServletException if a servlet-specific error occurs
     * @throws IOException if an I/O error occurs
     */
    @Override
    protected void doGet(HttpServletRequest request, HttpServletResponse response)
            throws ServletException, IOException {
        processRequest(request, response);
    }

    /**
     * Handles the HTTP <code>POST</code> method.
     *
     * @param request servlet request
     * @param response servlet response
     * @throws ServletException if a servlet-specific error occurs
     * @throws IOException if an I/O error occurs
     */
    @Override
    protected void doPost(HttpServletRequest request, HttpServletResponse response)
            throws ServletException, IOException {
        processRequest(request, response);
    }

    /**
     * Returns a short description of the servlet.
     *
     * @return a String containing servlet description
     */
    @Override
    public String getServletInfo() {
        return "Short description";
    }// </editor-fold>

}
```

AddLoginServlet.java

```
package cn.edu.djtu;

import java.io.IOException;
import java.io.PrintWriter;
import java.sql.Connection;
import java.sql.DriverManager;
import java.sql.PreparedStatement;
```

```
import java.sql.ResultSet;
import java.sql.SQLException;
import java.util.logging.Level;
import java.util.logging.Logger;
import javax.servlet.ServletException;
import javax.servlet.annotation.WebServlet;
import javax.servlet.http.HttpServlet;
import javax.servlet.http.HttpServletRequest;
import javax.servlet.http.HttpServletResponse;

@WebServlet(name = "AddLoginServlet", urlPatterns = {"/AddLoginServlet"})
public class AddLoginServlet extends HttpServlet {

    /**
     * Processes requests for both HTTP <code>GET</code> and <code>POST</code>
     * methods.
     *
     * @param request servlet request
     * @param response servlet response
     * @throws ServletException if a servlet-specific error occurs
     * @throws IOException if an I/O error occurs
     */
    protected void processRequest(HttpServletRequest request, HttpServletResponse response)
            throws ServletException, IOException {
        response.setContentType("text/html;charset=UTF-8");
        String username = request.getParameter("username");
        String userpass = request.getParameter("userpass");

        //连接数据库用到的对象
        Connection conn = null;
        PreparedStatement prst = null;
        ResultSet rs = null;

        //连接数据库用到的参数信息
        String url = "jdbc:derby://localhost:1527/sample";
        String driver = "org.apache.derby.jdbc.ClientDriver";
        String user = "app";
        String password = "app";

        //查询数据库的SQL语句
        String sql1 = "SELECT username , userpass FROM login WHERE username = ?";
        //添加用户的SQL语句
        String sql2 = "INSERT INTO login VALUES (?,?)";
        try (PrintWriter out = response.getWriter()) {
             Class.forName(driver);
            conn = DriverManager.getConnection(url, user, password);
            prst = conn.prepareStatement(sql1);
            prst.setString(1, username);
            rs = prst.executeQuery();
```

```
                if (rs.next()) {
                    request.setAttribute("error", "该用户已存在");
                    request.getRequestDispatcher("addLogin.jsp").forward(request, response);
                }else{
                    prst = conn.prepareStatement(sql2);
                    prst.setString(1, username);
                    prst.setString(2, userpass);
                    prst.executeUpdate();
                    request.getRequestDispatcher("index.jsp").forward(request, response);
                }
            } catch (ClassNotFoundException | SQLException ex) {
Logger.getLogger(AddLoginServlet.class.getName()).log(Level.SEVERE, null, ex);
            }finally{
                //关闭结果集对象
                if (rs != null) {
                    try {
                        rs.close();
                    } catch (SQLException ex) {
Logger.getLogger(AddLoginServlet.class.getName()).log(Level.SEVERE, null, ex);
                    }
                }
                //关闭预编译语句对象
                if (prst != null) {
                    try {
                        prst.close();
                    } catch (SQLException ex) {
Logger.getLogger(AddLoginServlet.class.getName()).log(Level.SEVERE, null, ex);
                    }
                }
                //关闭连接对象
                if (conn != null) {
                    try {
                        conn.close();
                    } catch (SQLException ex) {
Logger.getLogger(AddLoginServlet.class.getName()).log(Level.SEVERE, null, ex);
                    }
                }
            }
        }

        // <editor-fold defaultstate="collapsed" desc="HttpServlet methods. Click on the +
sign on the left to edit the code.">
        /**
         * Handles the HTTP <code>GET</code> method.
         *
         * @param request servlet request
         * @param response servlet response
         * @throws ServletException if a servlet-specific error occurs
         * @throws IOException if an I/O error occurs
```

```
     */
    @Override
    protected void doGet(HttpServletRequest request, HttpServletResponse response)
            throws ServletException, IOException {
        processRequest(request, response);
    }

    /**
     * Handles the HTTP <code>POST</code> method.
     *
     * @param request servlet request
     * @param response servlet response
     * @throws ServletException if a servlet-specific error occurs
     * @throws IOException if an I/O error occurs
     */
    @Override
    protected void doPost(HttpServletRequest request, HttpServletResponse response)
            throws ServletException, IOException {
        processRequest(request, response);
    }

    /**
     * Returns a short description of the servlet.
     *
     * @return a String containing servlet description
     */
    @Override
    public String getServletInfo() {
        return "Short description";
    }// </editor-fold>

}
```

10.4　MIS 第 2 版实现（添加数据库连接类）

对于 MIS 第 1 版的实现，仔细观察一下 LoginHandleServlet.java 和 AddLoginServlet.java 就会发现，在这两个 Servlet 中，有着完全相同的一部分代码，例如：

```
//连接数据库用到的参数信息
String url = "jdbc:derby://localhost:1527/sample";
String driver = "org.apache.derby.jdbc.ClientDriver";
String user = "app";
String password = "app";
```

这部分代码是数据库操作中的一些参数信息。下面的代码：

```
Class.forName(driver);
conn = DriverManager.getConnection(url, user, password);
```

这部分代码用于获得访问数据库的连接对象。还有关闭连接对象的代码：

```
//关闭连接对象
    if (conn != null) {
        try {
            conn.close();
        } catch (SQLException ex) {
Logger.getLogger(LoginHandleServlet.class.getName()).log(Level.SEVERE, null, ex);
        }
    }
```

那么能不能把这些反复出现的，基本完全相同的代码封装在一起呢？下面来实现 MIS 的第 2 个版本。添加一个对数据库连接进行操作的类，通过该类来完成创建连接对象以及关闭连接对象的任务。

注意：关闭结果集对象和预编译语句对象的代码虽然也重复了，但是不适合封装进工具类，因为操作对象不是相同的数据结构。

工程名：eg1002。

用到的文件如表 10.3 所示。

表 10.3 eg1002 用到的文件及文件说明

文件名	说明
login.jsp	用户登录页
index.jsp	首页
LoginHandleServlet.java	处理用户登录的 Servlet，位于 cn.edu.djtu 包中
addLogin.jsp	添加新用户的页面
AddLoginServlet.java	处理添加用户的 Servlet，位于 cn.edu.djtu 包中
DBConnection.java	用来创建连接对象及关闭连接对象的类，位于 cn.edu.djtu.util 包中

打开 NetBeans IDE，右击项目 eg1001，选择“复制”，在弹出的窗口中将项目名称设置为 eg1002，具体过程略。新建名为 DBConnection.java 的类，该类位于 cn.edu.djtu.util 包中。因为连接数据库需要的参数不再改变，所以为静态常量。编写代码如下：

DBConnection.java

```
package cn.edu.djtu.util;

import java.sql.Connection;
import java.sql.SQLException;
import java.util.logging.Level;
import java.util.logging.Logger;

public class DBConnection {
```

```
    private static final String URL = "jdbc:derby://localhost:1527/sample";
    private static final String DRIVER = "org.apache.derby.jdbc.ClientDriver";
    private static final String USER = "app";
    private static final String PASSWORD = "app";
    private Connection conn = null;

    public DBConnection(Connection conn) {
        try {
            Class.forName(DRIVER);
            conn = DriverManager.getConnection(URL, USER, PASSWORD);
        } catch (ClassNotFoundException | SQLException ex) {
        Logger.getLogger(DBConnection.class.getName()).log(Level.SEVERE, null, ex);
        }
    }

    public Connection getConnection(){
        return conn;
    }

    public void close(){
        try {
            conn.close();
        } catch (SQLException ex) {
    Logger.getLogger(DBConnection.class.getName()).log(Level.SEVERE, null, ex);
        }
    }
}
```

然后修改 LoginHandleServlet.java 代码,修订的部分已用波浪线标出。

LoginHandleServlet.java

```
package cn.edu.djtu;

import java.io.IOException;
import java.io.PrintWriter;
import java.sql.Connection;
import java.sql.DriverManager;
import java.sql.PreparedStatement;
import java.sql.ResultSet;
import java.sql.SQLException;
import java.util.logging.Level;
import java.util.logging.Logger;
import javax.servlet.ServletException;
import javax.servlet.annotation.WebServlet;
import javax.servlet.http.HttpServlet;
import javax.servlet.http.HttpServletRequest;
import javax.servlet.http.HttpServletResponse;
import javax.servlet.http.HttpSession;
```

```
@WebServlet(name = "LoginHandleServlet", urlPatterns = {"/LoginHandleServlet"})
public class LoginHandleServlet extends HttpServlet {

    /**
     * Processes requests for both HTTP <code>GET</code> and <code>POST</code>
     * methods.
     *
     * @param request servlet request
     * @param response servlet response
     * @throws ServletException if a servlet-specific error occurs
     * @throws IOException if an I/O error occurs
     */
    protected void processRequest(HttpServletRequest request, HttpServletResponse response)
            throws ServletException, IOException {
        response.setContentType("text/html;charset=UTF-8");
        String username = request.getParameter("username");
        String userpass = request.getParameter("userpass");

        //连接数据库用到的对象
        Connection conn = null;
        PreparedStatement prst = null;
        ResultSet rs = null;

        //实例化一个DBConnection对象
        DBConnection dbc = new DBConnection();

        //查询数据库的SQL语句
        String sql = "SELECT username , userpass FROM login WHERE username = ?";
        try (PrintWriter out = response.getWriter()) {
            conn = dbc.getConnection();
            prst = conn.prepareStatement(sql);
            prst.setString(1, username);
            rs = prst.executeQuery();
            if (rs.next() && rs.getString(2).equals(userpass)) {
                HttpSession session = request.getSession();
                session.setAttribute("username",username);
                request.getRequestDispatcher("index.jsp").forward(request, response);
            } else {
                request.setAttribute("error", "用户名密码错误");
                request.getRequestDispatcher("login.jsp").forward(request, response);
            }
        } catch (SQLException ex) {
Logger.getLogger(LoginHandleServlet.class.getName()).log(Level.SEVERE, null, ex);
        }finally{
            //关闭结果集对象
            if (rs != null) {
                try {
                    rs.close();
                } catch (SQLException ex) {
```

```
Logger.getLogger(LoginHandleServlet.class.getName()).log(Level.SEVERE, null, ex);
                    }
                }
                //关闭预编译语句对象
                if (prst != null) {
                    try {
                        prst.close();
                    } catch (SQLException ex) {
Logger.getLogger(LoginHandleServlet.class.getName()).log(Level.SEVERE, null, ex);
                    }
                }
                //关闭连接对象
                dbc.close();
            }
        }

    // <editor-fold defaultstate="collapsed" desc="HttpServlet methods. Click on the +
sign on the left to edit the code.">
    /**
     * Handles the HTTP <code>GET</code> method.
     *
     * @param request servlet request
     * @param response servlet response
     * @throws ServletException if a servlet-specific error occurs
     * @throws IOException if an I/O error occurs
     */
    @Override
    protected void doGet(HttpServletRequest request, HttpServletResponse response)
            throws ServletException, IOException {
        processRequest(request, response);
    }

    /**
     * Handles the HTTP <code>POST</code> method.
     *
     * @param request servlet request
     * @param response servlet response
     * @throws ServletException if a servlet-specific error occurs
     * @throws IOException if an I/O error occurs
     */
    @Override
    protected void doPost(HttpServletRequest request, HttpServletResponse response)
            throws ServletException, IOException {
        processRequest(request, response);
    }

    /**
     * Returns a short description of the servlet.
     *
```

```
     * @return a String containing servlet description
     */
    @Override
    public String getServletInfo() {
        return "Short description";
    }// </editor-fold>

}
```

然后修改 AddLoginServlet.java 代码,修订的部分已用波浪线标出。

AddLoginServlet.java

```
package cn.edu.djtu;

import java.io.IOException;
import java.io.PrintWriter;
import java.sql.Connection;
import java.sql.DriverManager;
import java.sql.PreparedStatement;
import java.sql.ResultSet;
import java.sql.SQLException;
import java.util.logging.Level;
import java.util.logging.Logger;
import javax.servlet.ServletException;
import javax.servlet.annotation.WebServlet;
import javax.servlet.http.HttpServlet;
import javax.servlet.http.HttpServletRequest;
import javax.servlet.http.HttpServletResponse;

@WebServlet(name = "AddLoginServlet", urlPatterns = {"/AddLoginServlet"})
public class AddLoginServlet extends HttpServlet {

    /**
     * Processes requests for both HTTP <code>GET</code> and <code>POST</code>
     * methods.
     *
     * @param request servlet request
     * @param response servlet response
     * @throws ServletException if a servlet-specific error occurs
     * @throws IOException if an I/O error occurs
     */
    protected void processRequest(HttpServletRequest request, HttpServletResponse response)
            throws ServletException, IOException {
        response.setContentType("text/html;charset=UTF-8");
        String username = request.getParameter("username");
        String userpass = request.getParameter("userpass");

        //连接数据库用到的对象
```

```
        Connection conn = null;
        PreparedStatement prst = null;
        ResultSet rs = null;

        //实例化一个 DBConnection 对象
        DBConnection dbc = new DBConnection();

        //查询数据库的 SQL 语句
        String sql1 = "SELECT username , userpass FROM login WHERE username = ?";
        //添加用户的 SQL 语句
        String sql2 = "INSERT INTO login VALUES (?,?)";
        try (PrintWriter out = response.getWriter()) {
            conn = dbc.getConnection();
            prst = conn.prepareStatement(sql1);
            prst.setString(1, username);
            rs = prst.executeQuery();
            if (rs.next()) {
                request.setAttribute("error", "该用户已存在");
                request.getRequestDispatcher("addLogin.jsp").forward(request, response);
            }else{
                prst = conn.prepareStatement(sql2);
                prst.setString(1, username);
                prst.setString(2, userpass);
                prst.executeUpdate();
                request.getRequestDispatcher("index.jsp").forward(request, response);
            }
        } catch (SQLException ex) {
Logger.getLogger(AddLoginServlet.class.getName()).log(Level.SEVERE, null, ex);
        }finally{
            //关闭结果集对象
            if (rs != null) {
                try {
                    rs.close();
                } catch (SQLException ex) {
Logger.getLogger(AddLoginServlet.class.getName()).log(Level.SEVERE, null, ex);
                }
            }
            //关闭预编译语句对象
            if (prst != null) {
                try {
                    prst.close();
                } catch (SQLException ex) {
Logger.getLogger(AddLoginServlet.class.getName()).log(Level.SEVERE, null, ex);
                }
            }
            //关闭连接对象
            dbc.close()
        }
    }
```

```
    // <editor-fold defaultstate="collapsed" desc="HttpServlet methods. Click on the +
sign on the left to edit the code.">
    /**
     * Handles the HTTP <code>GET</code> method.
     *
     * @param request servlet request
     * @param response servlet response
     * @throws ServletException if a servlet-specific error occurs
     * @throws IOException if an I/O error occurs
     */
    @Override
    protected void doGet(HttpServletRequest request, HttpServletResponse response)
            throws ServletException, IOException {
        processRequest(request, response);
    }

    /**
     * Handles the HTTP <code>POST</code> method.
     *
     * @param request servlet request
     * @param response servlet response
     * @throws ServletException if a servlet-specific error occurs
     * @throws IOException if an I/O error occurs
     */
    @Override
    protected void doPost(HttpServletRequest request, HttpServletResponse response)
            throws ServletException, IOException {
        processRequest(request, response);
    }

    /**
     * Returns a short description of the servlet.
     *
     * @return a String containing servlet description
     */
    @Override
    public String getServletInfo() {
        return "Short description";
    }// </editor-fold>

}
```

观察代码可以发现与数据库连接的参数部分在Servlet中不再出现，获得数据库连接和关闭数据库连接的操作都集中封装到了DBConnection中。这样做的好处就是当需要更换数据库时，只需要更改DBConnection中的代码即可。

同时也可以看到，Servlet中此时混杂了对数据库中数据的操作和对流程控制操作。想要彻底分离,就要采用分层的设计思想,将程序分成若干层,每一层实现特定的功能。在第7章的学习中,已经应用了这种思想做过简单的MVC设计模式的例子,接下来在此基础上更进一步,应用一下DAO设计模式,并且引入POJO。

10.5 MIS第3版实现(添加POJO与DAO)

1. POJO简介

POJO是Plain Ordinary Java Object的缩写,翻译为简单的Java对象。在使用中很多人可能会把POJO和JavaBean混为一谈,其实两者还是有区别的。

POJO肯定满足JavaBean的规范,但不是所有的JavaBean都满足POJO的规范。POJO结构比较单一,由属性和setter/getter方法组成,而JavaBean中除了这些以外还可以有其他业务代码。这个不同就决定了两者应用场合不完全一样。在Java Web开发中,曾经有过的模式一(JSP+JavaBean),其中的JavaBean往往包含着业务逻辑。而在模式二(JSP+Servlet+JavaBean)中的JavaBean,则基本上就可以看作是POJO。在第7章中曾经使用过的JavaBean,也是按照POJO的方式在使用。在当前的Java Web开发中,把POJO当作数据存储的载体。强调POJO的概念而淡化JavaBean的概念。

要编写一个POJO类,一般需要遵循如下的规范。

(1) POJO类名与数据库中的表名一一对应,例如:Customer类对应customer表。

(2) POJO类的属性通常与数据库中对应表的字段一一对应,属性的数据类型应为引用类型,而不是基本数据类型,例如:不使用int,而是使用Integer。

(3) POJO类的属性访问权限为private,提供访问权限为public的存取控制方法(setter/getter方法)。

(4) POJO类中一定要包含公共类型且不含参数的构造方法。

以上几点要求与JavaBean的要求类似。此外,POJO类还可能有的规范是:

(1) 为了便于程序扩展,POJO类可以实现java.io.Serializable序列化接口。

(2) POJO类可以根据需要重写Object类的:equals()、hashCode()和toString()方法。

(3) POJO类集中存放在名为vo或者bean的包中。

除了JavaBean以外,POJO还容易与一些概念混淆,例如:VO(Value Object,值对象)、DTO(Data Transfer Object,数据传输对象)、DO(Domain Object,域对象)、PO(Persistent Object,持久化对象)。PO是由ORM框架负责,暂时不需要考虑。DO与VO和DTO的主要差别在于DO的属性与数据库中表的字段要一一对应,既不多也不少。而DTO或者VO中的属性可以少于DTO,因为数据传递过程也好,视图层显示也好,可能只使用了表中部分字段。

POJO的命名一般也要体现与表的对应关系,以本章案例为例,与login表对应的POJO类一般命名为Login.java,放在cn.edu.djtu.vo或者cn.edu.djtu.bean包中。

2. DAO简介

POJO能做到的是将关系数据库中一条记录封装成一个对象,但是无法完成增、删、改、查这类操作。而 DAO 则是专门完成这类操作的对象。DAO(Data Access Object)被称为数据访问对象,封装了对数据的操作。

在已经完成的前两版程序中,登录需要按用户名查询 login 表,添加新用户也需要按用户名查询 login 表,这两次查询操作是没有差别的,那么就可以考虑把对 login 表的增、删、改、查之类的操作都封装在一起。

既然只封装了操作,那么就可以将 DAO 声明成接口,使用时则用该接口的实现类。命名时为了维护方便,有一定的要求,以本章案例为例:数据库中表的名字是 login,那么对应的 DAO 接口命名为 LoginDAO.java,接口的实现类命名为 LoginDAOImpl.java。LoginDAO.java 放在 cn.edu.djtu.dao 包中,LoginDAOImpl.java 放在 cn.edu.djtu.dao.impl 包中。

3. 第 3 版代码实现

打开 NetBeans IDE,右击项目 eg1002,选择“复制”命令,在弹出的窗口中将项目名称设置为 eg1003,具体过程略。

工程名:eg1003。

项目用到的文件如表 10.4 所示。

表 10.4 eg1003 用到的文件及文件说明

文件名	说明
login.jsp	用户登录页
index.jsp	首页
LoginHandleServlet.java	处理用户登录的 Servlet,位于 cn.edu.djtu 包中
addLogin.jsp	添加新用户的页面
AddLoginServlet.java	处理添加用户的 Servlet,位于 cn.edu.djtu 包中
DBConnection.java	用来创建连接对象及关闭连接对象的类,位于 cn.edu.djtu.util 包中
Login.java	与 login 表对应的 POJO
LoginDAO.java	与 login 表对应的 DAO 接口
LoginDAOImpl.java	与 login 表对应的 LoginDAO 接口实现类

新建名为 Login.java 的 POJO 类,包名 cn.edu.djtu.vo。编写代码如下,编写时可以通过封装字段的方式自动生成 setter/getter 方法。

Login.java

```
package cn.edu.djtu.vo;

public class Login {
    private String username;
    private String userpass;

    /**
     * @return the username
     */
    public String getUsername() {
```

```
        return username;
    }

    /**
     * @param username the username to set
     */
    public void setUsername(String username) {
        this.username = username;
    }

    /**
     * @return the userpass
     */
    public String getUserpass() {
        return userpass;
    }

    /**
     * @param userpass the userpass to set
     */
    public void setUserpass(String userpass) {
        this.userpass = userpass;
    }
}
```

编写 LoginDAO 接口时，如果是首次新建接口，那么右击后弹出的菜单中可能没有“Java 接口”命令，此时可以选择“其他”命令，如图 10.7 所示。

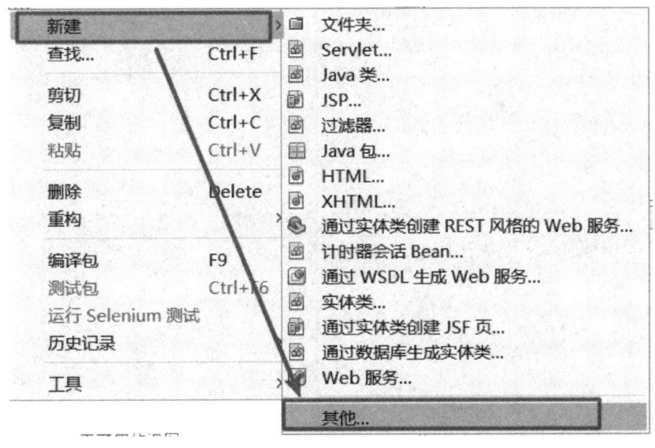

图 10.7 新建 Java 接口时，选择“其他”命令

然后在弹出的窗口中选择“类别”中的 Java，“文件类型”选择“Java 接口”，如图 10.8 所示。

第二次创建时，就可以直接选择右击后弹出的菜单中“Java 接口”的命令。

编写 LoginDAO.java 接口代码如下，此处只编写了与案例相关的两个方法。

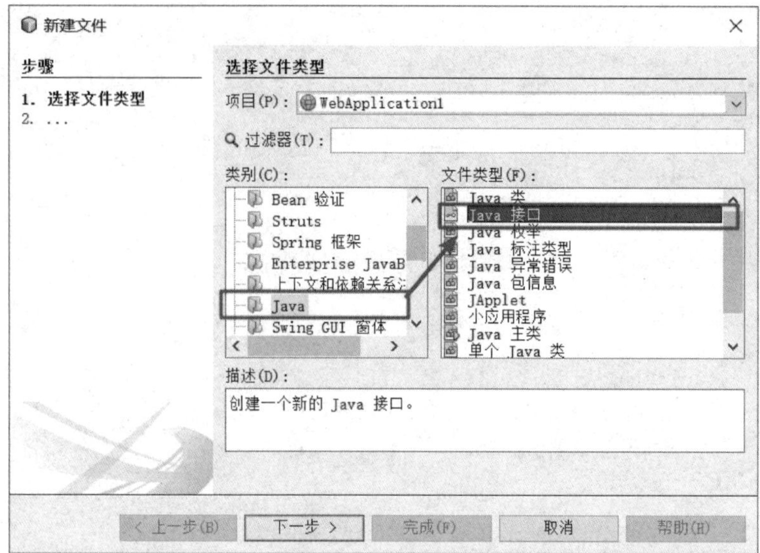

图 10.8 创建 Java 接口

LoginDAO.java

```
package cn.edu.djtu.dao;

import cn.edu.djtu.vo.Login;

public interface LoginDAO {
    public abstract Login findByName(String username) throws Exception;
    public abstract boolean save(Login login) throws Exception;
}
```

编写 LoginDAOImpl.java 类代码如下：

LoginDAOImpl.java

```
package cn.edu.djtu.dao.impl;

import cn.edu.djtu.dao.LoginDAO;
import cn.edu.djtu.vo.Login;
import java.sql.Connection;
import java.sql.PreparedStatement;
import java.sql.ResultSet;

public class LoginDAOImpl implements LoginDAO {

    private Connection conn = null;

    public LoginDAOImpl(Connection conn) {
        this.conn = conn;
```

```
    }

    @Override
    public Login findByName(String username) throws Exception {
        Login login = null;
        PreparedStatement prst;
        ResultSet rs;
        String sql = "SELECT username, userpass FROM login WHERE username = ?";
        prst = conn.prepareStatement(sql);
        prst.setString(1, username);
        rs = prst.executeQuery();
        if (rs.next()) {
            login = new Login();
            login.setUsername(rs.getString(1));
            login.setUserpass(rs.getString(2));
        }
        rs.close();
        prst.close();
        return login;
    }

    @Override
    public boolean save(Login login) throws Exception {
        boolean b = false;
        PreparedStatement prst;
        String sql = "INSERT INTO login VALUES(?,?)";
        prst = conn.prepareStatement(sql);
        prst.setString(1, login.getUsername());
        prst.setString(2, login.getUserpass());
        if (prst.executeUpdate() > 0) {
            b = true;
        }
        prst.close();
        return b;
    }
}
```

然后修改 LoginHandleServlet.java 代码。

LoginHandleServlet.java

```
package cn.edu.djtu;

import cn.edu.djtu.dao.LoginDAO;
import cn.edu.djtu.dao.impl.LoginDAOImpl;
import cn.edu.djtu.util.DBConnection;
import cn.edu.djtu.vo.Login;
import java.io.IOException;
import java.io.PrintWriter;
```

```
import java.util.logging.Level;
import java.util.logging.Logger;
import javax.servlet.ServletException;
import javax.servlet.annotation.WebServlet;
import javax.servlet.http.HttpServlet;
import javax.servlet.http.HttpServletRequest;
import javax.servlet.http.HttpServletResponse;
import javax.servlet.http.HttpSession;

@WebServlet(name = "LoginHandleServlet", urlPatterns = {"/LoginHandleServlet"})
public class LoginHandleServlet extends HttpServlet {

    /**
     * Processes requests for both HTTP <code>GET</code> and <code>POST</code>
     * methods.
     *
     * @param request servlet request
     * @param response servlet response
     * @throws ServletException if a servlet-specific error occurs
     * @throws IOException if an I/O error occurs
     */
    protected void processRequest(HttpServletRequest request, HttpServletResponse response)
            throws ServletException, IOException {
        response.setContentType("text/html;charset=UTF-8");
        String username = request.getParameter("username");
        String userpass = request.getParameter("userpass");

        //实例化一个 DBConnection 对象
        DBConnection dbc = new DBConnection();
        //实例化一个 LoginDAOImpl
        LoginDAO loginDAO = new LoginDAOImpl(dbc.getConnection());

        try (PrintWriter out = response.getWriter()) {
            Login login = loginDAO.findByName(username);
            if (login != null && login.getUserpass().equals(userpass)) {
                HttpSession session = request.getSession();
                session.setAttribute("username", username);
                request.getRequestDispatcher("index.jsp").forward(request, response);
            } else {
                request.setAttribute("error", "用户名密码错误");
                request.getRequestDispatcher("login.jsp").forward(request, response);
            }
        } catch (Exception ex) {
Logger.getLogger(LoginHandleServlet.class.getName()).log(Level.SEVERE, null, ex);
        } finally {
            //关闭连接对象
            dbc.close();
        }
    }
```

```
    // <editor-fold defaultstate="collapsed" desc="HttpServlet methods. Click on the + sign on the left to edit the code.">
    /**
     * Handles the HTTP <code>GET</code> method.
     *
     * @param request servlet request
     * @param response servlet response
     * @throws ServletException if a servlet-specific error occurs
     * @throws IOException if an I/O error occurs
     */
    @Override
    protected void doGet(HttpServletRequest request, HttpServletResponse response)
            throws ServletException, IOException {
        processRequest(request, response);
    }

    /**
     * Handles the HTTP <code>POST</code> method.
     *
     * @param request servlet request
     * @param response servlet response
     * @throws ServletException if a servlet-specific error occurs
     * @throws IOException if an I/O error occurs
     */
    @Override
    protected void doPost(HttpServletRequest request, HttpServletResponse response)
            throws ServletException, IOException {
        processRequest(request, response);
    }

    /**
     * Returns a short description of the servlet.
     *
     * @return a String containing servlet description
     */
    @Override
    public String getServletInfo() {
        return "Short description";
    }// </editor-fold>

}
```

然后修改 AddLoginServlet.java 代码。

AddLoginServlet.java

```
package cn.edu.djtu;

import cn.edu.djtu.dao.LoginDAO;
```

```
import cn.edu.djtu.dao.impl.LoginDAOImpl;
import cn.edu.djtu.util.DBConnection;
import cn.edu.djtu.vo.Login;
import java.io.IOException;
import java.io.PrintWriter;
import java.util.logging.Level;
import java.util.logging.Logger;
import javax.servlet.ServletException;
import javax.servlet.annotation.WebServlet;
import javax.servlet.http.HttpServlet;
import javax.servlet.http.HttpServletRequest;
import javax.servlet.http.HttpServletResponse;

@WebServlet(name = "AddLoginServlet", urlPatterns = {"/AddLoginServlet"})
public class AddLoginServlet extends HttpServlet {

    /**
     * Processes requests for both HTTP <code>GET</code> and <code>POST</code>
     * methods.
     *
     * @param request servlet request
     * @param response servlet response
     * @throws ServletException if a servlet-specific error occurs
     * @throws IOException if an I/O error occurs
     */
    protected void processRequest(HttpServletRequest request, HttpServletResponse response)
            throws ServletException, IOException {
        response.setContentType("text/html;charset=UTF-8");
        String username = request.getParameter("username");
        String userpass = request.getParameter("userpass");

        //实例化一个DBConnection对象
        DBConnection dbc = new DBConnection();
        //实例化一个LoginDAOImpl
        LoginDAO loginDAO = new LoginDAOImpl(dbc.getConnection());

        try (PrintWriter out = response.getWriter()) {
            if (loginDAO.findByName(username) != null) {
                request.setAttribute("error", "该用户已存在");
                request.getRequestDispatcher("addLogin.jsp").forward(request, response);
            } else {
                Login login = new Login();
                login.setUsername(username);
                login.setUserpass(userpass);
                loginDAO.save(login);
                request.getRequestDispatcher("index.jsp").forward(request, response);
            }
        }catch (Exception ex) {
Logger.getLogger(AddLoginServlet.class.getName()).log(Level.SEVERE, null, ex);
```

```
        } finally {
            //关闭连接对象
            dbc.close();
        }
    }

    // <editor-fold defaultstate="collapsed" desc="HttpServlet methods. Click on the + sign on the left to edit the code.">
    /**
     * Handles the HTTP <code>GET</code> method.
     *
     * @param request servlet request
     * @param response servlet response
     * @throws ServletException if a servlet-specific error occurs
     * @throws IOException if an I/O error occurs
     */
    @Override
    protected void doGet(HttpServletRequest request, HttpServletResponse response)
            throws ServletException, IOException {
        processRequest(request, response);
    }

    /**
     * Handles the HTTP <code>POST</code> method.
     *
     * @param request servlet request
     * @param response servlet response
     * @throws ServletException if a servlet-specific error occurs
     * @throws IOException if an I/O error occurs
     */
    @Override
    protected void doPost(HttpServletRequest request, HttpServletResponse response)
            throws ServletException, IOException {
        processRequest(request, response);
    }

    /**
     * Returns a short description of the servlet.
     *
     * @return a String containing servlet description
     */
    @Override
    public String getServletInfo() {
        return "Short description";
    }// </editor-fold>

}
```

其他文件代码不变。由 Servlet 中的代码可见，Servlet 中已经没有具体的数据库操作的代码了。数据库操作相关的代码都被封装到了 DAO 层。

10.6 MIS 第 4 版实现(添加 DAO 工厂)

在第 3 版的实现中，需要在 Servlet 中实例化 LoginDAOImpl，语句为 LoginDAO loginDAO = new LoginDAOImpl(dbc.getConnection())；。这使得充当控制器的 Servlet 和 DAO 层耦合在了一起，此时可以考虑采用工厂模式来解耦合。

创建一个 DAO 工厂类，该类生产各种各样的 DAO 实例。在 Servlet 中需要 DAO 实例时，直接从工厂取一个即可。

打开 NetBeans IDE，右击项目 eg1003，在弹出的窗口中将项目名称设置为 eg1004，具体过程略。

工程名：eg1004。

项目用到的文件如表 10.5 所示。

表 10.5　eg1004 用到的文件及文件说明

文　件　名	说　　明
login.jsp	用户登录页
index.jsp	首页
LoginHandleServlet.java	处理用户登录的 Servlet，位于 cn.edu.djtu 包中
addLogin.jsp	添加新用户的页面
AddLoginServlet.java	处理添加用户的 Servlet，位于 cn.edu.djtu 包中
DBConnection.java	用来创建连接对象及关闭连接对象的类，位于 cn.edu.djtu.util 包中
Login.java	与 login 表对应的 POJO
LoginDAO.java	与 login 表对应的 DAO 接口
LoginDAOImpl.java	与 login 表对应的 LoginDAO 接口实现类
DAOFactory.java	DAO 工厂类，产生各种 DAO 实例

新建名为 DAOFactory.java 的 DAO 工厂类，包名是 cn.edu.djtu.factory。编写代码如下：

DAOFactory.java

```
package cn.edu.djtu.factory;

import cn.edu.djtu.dao.impl.LoginDAOImpl;
import java.sql.Connection;

public class DAOFactory {
    public static LoginDAOImpl getLoginDAOImpl(Connection conn){
        return new LoginDAOImpl(conn);
    }
}
```

然后修改 LoginHandleServlet.java 代码。

LoginHandleServlet.java

```
package cn.edu.djtu;

import cn.edu.djtu.dao.LoginDAO;
import cn.edu.djtu.util.DBConnection;
import cn.edu.djtu.vo.Login;
import java.io.IOException;
import java.io.PrintWriter;
import java.util.logging.Level;
import java.util.logging.Logger;
import javax.servlet.ServletException;
import javax.servlet.annotation.WebServlet;
import javax.servlet.http.HttpServlet;
import javax.servlet.http.HttpServletRequest;
import javax.servlet.http.HttpServletResponse;
import javax.servlet.http.HttpSession;

@WebServlet(name = "LoginHandleServlet", urlPatterns = {"/LoginHandleServlet"})
public class LoginHandleServlet extends HttpServlet {

    /**
     * Processes requests for both HTTP <code>GET</code> and <code>POST</code>
     * methods.
     *
     * @param request servlet request
     * @param response servlet response
     * @throws ServletException if a servlet-specific error occurs
     * @throws IOException if an I/O error occurs
     */
    protected void processRequest(HttpServletRequest request, HttpServletResponse response)
            throws ServletException, IOException {
        response.setContentType("text/html;charset=UTF-8");
        String username = request.getParameter("username");
        String userpass = request.getParameter("userpass");

        //实例化一个 DBConnection 对象
        DBConnection dbc = new DBConnection();
        //通过工厂类获得一个 LoginDAOImpl 实例
        LoginDAO loginDAO = DAOFactory.getLoginDAOImpl(dbc.getConnection());

        try (PrintWriter out = response.getWriter()) {
            Login login = loginDAO.findByName(username);
            if (login != null && login.getUserpass().equals(userpass)) {
                HttpSession session = request.getSession();
                session.setAttribute("username", username);
                request.getRequestDispatcher("index.jsp").forward(request, response);
            } else {
```

```
                request.setAttribute("error", "用户名密码错误");
                request.getRequestDispatcher("login.jsp").forward(request, response);
            }
        } catch (Exception ex) {
Logger.getLogger(LoginHandleServlet.class.getName()).log(Level.SEVERE, null, ex);
        } finally {
            //关闭连接对象
            dbc.close();
        }
    }

    // <editor-fold defaultstate="collapsed" desc="HttpServlet methods. Click on the +
sign on the left to edit the code.">
    /**.
     * Handles the HTTP <code>GET</code> method.
     *
     * @param request servlet request
     * @param response servlet response
     * @throws ServletException if a servlet-specific error occurs
     * @throws IOException if an I/O error occurs
     */
    @Override
    protected void doGet(HttpServletRequest request, HttpServletResponse response)
            throws ServletException, IOException {
        processRequest(request, response);
    }

    /**
     * Handles the HTTP <code>POST</code> method.
     *
     * @param request servlet request
     * @param response servlet response
     * @throws ServletException if a servlet-specific error occurs
     * @throws IOException if an I/O error occurs
     */
    @Override
    protected void doPost(HttpServletRequest request, HttpServletResponse response)
            throws ServletException, IOException {
        processRequest(request, response);
    }

    /**
     * Returns a short description of the servlet.
     *
     * @return a String containing servlet description
     */
    @Override
    public String getServletInfo() {
        return "Short description";
    }// </editor-fold>

}
```

然后修改 AddLoginServlet.java 代码。

AddLoginServlet.java

```
package cn.edu.djtu;

import cn.edu.djtu.dao.LoginDAO;
import cn.edu.djtu.util.DBConnection;
import cn.edu.djtu.vo.Login;
import java.io.IOException;
import java.io.PrintWriter;
import java.util.logging.Level;
import java.util.logging.Logger;
import javax.servlet.ServletException;
import javax.servlet.annotation.WebServlet;
import javax.servlet.http.HttpServlet;
import javax.servlet.http.HttpServletRequest;
import javax.servlet.http.HttpServletResponse;

@WebServlet(name = "AddLoginServlet", urlPatterns = {"/AddLoginServlet"})
public class AddLoginServlet extends HttpServlet {

    /**
     * Processes requests for both HTTP <code>GET</code> and <code>POST</code>
     * methods.
     *
     * @param request servlet request
     * @param response servlet response
     * @throws ServletException if a servlet-specific error occurs
     * @throws IOException if an I/O error occurs
     */
    protected void processRequest(HttpServletRequest request, HttpServletResponse response)
            throws ServletException, IOException {
        response.setContentType("text/html;charset=UTF-8");
        String username = request.getParameter("username");
        String userpass = request.getParameter("userpass");

        //实例化一个 DBConnection 对象
        DBConnection dbc = new DBConnection();
        //通过工厂类获得一个 LoginDAOImpl 实例
        LoginDAO loginDAO = DAOFactory.getLoginDAOImpl(dbc.getConnection());

        try (PrintWriter out = response.getWriter()) {
            if (loginDAO.findByName(username) != null) {
                request.setAttribute("error", "该用户已存在");
                request.getRequestDispatcher("addLogin.jsp").forward(request, response);
            } else {
                Login login = new Login();
                login.setUsername(username);
                login.setUserpass(userpass);
```

```
                loginDAO.save(login);
                request.getRequestDispatcher("index.jsp").forward(request, response);
            }
        }catch (Exception ex) {
Logger.getLogger(AddLoginServlet.class.getName()).log(Level.SEVERE, null, ex);
        } finally {
            //关闭连接对象
            dbc.close();
        }
    }

    // <editor-fold defaultstate="collapsed" desc="HttpServlet methods. Click on the +
sign on the left to edit the code.">
    /**
     * Handles the HTTP <code>GET</code> method.
     *
     * @param request servlet request
     * @param response servlet response
     * @throws ServletException if a servlet-specific error occurs
     * @throws IOException if an I/O error occurs
     */
    @Override
    protected void doGet(HttpServletRequest request, HttpServletResponse response)
            throws ServletException, IOException {
        processRequest(request, response);
    }

    /**
     * Handles the HTTP <code>POST</code> method.
     *
     * @param request servlet request
     * @param response servlet response
     * @throws ServletException if a servlet-specific error occurs
     * @throws IOException if an I/O error occurs
     */
    @Override
    protected void doPost(HttpServletRequest request, HttpServletResponse response)
            throws ServletException, IOException {
        processRequest(request, response);
    }

    /**
     * Returns a short description of the servlet.
     *
     * @return a String containing servlet description
     */
    @Override
    public String getServletInfo() {
        return "Short description";
    }// </editor-fold>

}
```

10.7　MIS 第 5 版实现(添加 Service 及 Service 工厂)

在第 4 版实现中,采用了 DAO 设计模式,对于一些简单应用,这种程度的层次划分已经可以了。但是对于业务复杂的情况,一般会进一步分层。例如,如果一个业务涉及多个 DAO 层的操作,那么可以考虑封装成一个 Service(服务)。与 DAO 层的结构类似,也包括 Service 工厂类、Service 接口和 Service 接口的实现类。

打开 NetBeans IDE,右击项目 eg1004,选择"复制"命令,在弹出的窗口中将项目名称设置为 eg1005,具体过程略。

工程名:eg1005。

项目用到的文件如表 10.6 所示。

表 10.6　eg1005 用到的文件及文件说明

文　件　名	说　　明
login.jsp	用户登录页
index.jsp	首页
LoginHandleServlet.java	处理用户登录的 Servlet,位于 cn.edu.djtu 包中
addLogin.jsp	添加新用户的页面
AddLoginServlet.java	处理添加用户的 Servlet,位于 cn.edu.djtu 包中
DBConnection.java	用来创建连接对象及关闭连接对象的类,位于 cn.edu.djtu.util 包中
Login.java	与 login 表对应的 POJO
LoginDAO.java	与 login 表对应的 DAO 接口
LoginDAOImpl.java	与 login 表对应的 LoginDAO 接口实现类
DAOFactory.java	DAO 工厂类,产生各种 DAO 实例
LoginService.java	与 login 表相关的 Service 接口
LoginServiceImpl.java	与 login 表对应的 LoginService 接口实现类
ServiceFactory.java	Service 工厂类,产生各种 Service 实例

编写 LoginService.java 接口代码如下,此处只编写了与案例相关的两个方法。Service 接口中的方法,命名时能见其名知其意即可,以下只声明了两个与案例相关的服务——登录服务 logon 和添加登录用户的 addLogin。

LoginService.java

```
package cn.edu.djtu.service;

import cn.edu.djtu.vo.Login;

public interface LoginService {
    public abstract String logon(Login login) ;
    public abstract String addLogin(Login login);
}
```

编写 LoginServiceImpl. java 类代码如下：

LoginServiceImpl. java

```
package cn.edu.djtu.service.impl;

import cn.edu.djtu.dao.LoginDAO;
import cn.edu.djtu.factory.DAOFactory;
import cn.edu.djtu.service.LoginService;
import cn.edu.djtu.util.DBConnection;
import cn.edu.djtu.vo.Login;
import java.sql.Connection;
import java.util.logging.Level;
import java.util.logging.Logger;

public class LoginServiceImpl implements LoginService {

    private final DBConnection dbc = new DBConnection();

    @Override
    public String logon(Login login) {
        String string = "error";
        Connection conn = dbc.getConnection();
        LoginDAO loginDAO = DAOFactory.getLoginDAOImpl(conn);
        try {
            Login login1 = loginDAO.findByName(login.getUsername());
            if (login1 != null && login1.getUserpass().equals(login.getUserpass())) {
                string = "success";
            }
        } catch (Exception ex) {
Logger.getLogger(LoginServiceImpl.class.getName()).log(Level.SEVERE, null, ex);
        } finally {
            dbc.close();
        }
        return string;
    }

    @Override
    public String addLogin(Login login) {
        String string = "error";
        Connection conn = dbc.getConnection();
        LoginDAO loginDAO = DAOFactory.getLoginDAOImpl(conn);
        try {
              if (loginDAO.findByName(login.getUsername()) == null && loginDAO.save
(login)) {
                string = "success";
            }
        } catch (Exception ex) {
Logger.getLogger(LoginServiceImpl.class.getName()).log(Level.SEVERE, null, ex);
        } finally {
```

```
            dbc.close();
        }
        return string;
    }

}
```

新建名为 ServiceFactory.java 的 Service 工厂类，包名 cn.edu.djtu.factory。编写代码如下：

ServiceFactory.java

```
package cn.edu.djtu.factory;

import cn.edu.djtu.service.impl.LoginServiceImpl;

public class ServiceFactory {

    public static LoginServiceImpl getLoginServiceImpl() {
        return new LoginServiceImpl();
    }
}
```

然后修改 LoginHandleServlet.java 代码。

LoginHandleServlet.java

```
package cn.edu.djtu;

import cn.edu.djtu.factory.ServiceFactory;
import cn.edu.djtu.service.LoginService;
import cn.edu.djtu.vo.Login;
import java.io.IOException;
import javax.servlet.ServletException;
import javax.servlet.annotation.WebServlet;
import javax.servlet.http.HttpServlet;
import javax.servlet.http.HttpServletRequest;
import javax.servlet.http.HttpServletResponse;
import javax.servlet.http.HttpSession;

@WebServlet(name = "LoginHandleServlet", urlPatterns = {"/LoginHandleServlet"})
public class LoginHandleServlet extends HttpServlet {

    /**
     * Processes requests for both HTTP <code>GET</code> and <code>POST</code>
     * methods.
     *
     * @param request servlet request
     * @param response servlet response
```

```
     * @throws ServletException if a servlet - specific error occurs
     * @throws IOException if an I/O error occurs
     */
    protected void processRequest(HttpServletRequest request, HttpServletResponse response)
            throws ServletException, IOException {
        response.setContentType("text/html;charset = UTF - 8");
        String username = request.getParameter("username");
        String userpass = request.getParameter("userpass");

        Login login = new Login();
        login.setUsername(username);
        login.setUserpass(userpass);

        //通过工厂类获得一个 LoginServiceImpl 实例
    LoginService loginService = ServiceFactory.getLoginServiceImpl();

        if (loginService.logon(login).equals("success")) {
            HttpSession session = request.getSession();
            session.setAttribute("username", username);
request.getRequestDispatcher("index.jsp").forward(request, response);
        } else {
            request.setAttribute("error", "用户名密码错误");
request.getRequestDispatcher("login.jsp").forward(request, response);
        }
    }

    // <editor - fold defaultstate = "collapsed" desc = "HttpServlet methods. Click on the +
sign on the left to edit the code.">
    /**
     * Handles the HTTP <code> GET </code> method.
     *
     * @param request servlet request
     * @param response servlet response
     * @throws ServletException if a servlet - specific error occurs
     * @throws IOException if an I/O error occurs
     */
    @Override
    protected void doGet(HttpServletRequest request, HttpServletResponse response)
            throws ServletException, IOException {
        processRequest(request, response);
    }

    /**
     * Handles the HTTP <code> POST </code> method.
     *
     * @param request servlet request
     * @param response servlet response
     * @throws ServletException if a servlet - specific error occurs
     * @throws IOException if an I/O error occurs
```

```
     */
    @Override
    protected void doPost(HttpServletRequest request, HttpServletResponse response)
            throws ServletException, IOException {
        processRequest(request, response);
    }

    /**
     * Returns a short description of the servlet.
     *
     * @return a String containing servlet description
     */
    @Override
    public String getServletInfo() {
        return "Short description";
    }// </editor-fold>

}
```

然后修改 AddLoginServlet.java 代码。

AddLoginServlet.java

```
package cn.edu.djtu;

import cn.edu.djtu.factory.ServiceFactory;
import cn.edu.djtu.service.LoginService;
import cn.edu.djtu.vo.Login;
import java.io.IOException;
import javax.servlet.ServletException;
import javax.servlet.annotation.WebServlet;
import javax.servlet.http.HttpServlet;
import javax.servlet.http.HttpServletRequest;
import javax.servlet.http.HttpServletResponse;

@WebServlet(name = "AddLoginServlet", urlPatterns = {"/AddLoginServlet"})
public class AddLoginServlet extends HttpServlet {

    /**
     * Processes requests for both HTTP <code>GET</code> and <code>POST</code>
     * methods.
     *
     * @param request servlet request
     * @param response servlet response
     * @throws ServletException if a servlet-specific error occurs
     * @throws IOException if an I/O error occurs
     */
    protected void processRequest(HttpServletRequest request, HttpServletResponse response)
            throws ServletException, IOException {
```

```
        response.setContentType("text/html;charset = UTF - 8");
        String username = request.getParameter("username");
        String userpass = request.getParameter("userpass");

        Login login = new Login();
        login.setUsername(username);
        login.setUserpass(userpass);

        //通过工厂类获得一个 LoginServiceImpl 实例
    LoginService loginService = ServiceFactory.getLoginServiceImpl();

        if (loginService.addLogin(login).equals("error")) {
            request.setAttribute("error", "该用户已存在");
   request.getRequestDispatcher("addLogin.jsp").forward(request, response);
        } else {
request.getRequestDispatcher("index.jsp").forward(request, response);
        }

    }

    // <editor - fold defaultstate = "collapsed" desc = "HttpServlet methods. Click on the +
sign on the left to edit the code.">
    /**
     * Handles the HTTP <code> GET </code> method.
     *
     * @param request servlet request
     * @param response servlet response
     * @throws ServletException if a servlet - specific error occurs
     * @throws IOException if an I/O error occurs
     */
    @Override
    protected void doGet(HttpServletRequest request, HttpServletResponse response)
            throws ServletException, IOException {
        processRequest(request, response);
    }

    /**
     * Handles the HTTP <code> POST </code> method.
     *
     * @param request servlet request
     * @param response servlet response
     * @throws ServletException if a servlet - specific error occurs
     * @throws IOException if an I/O error occurs
     */
    @Override
    protected void doPost(HttpServletRequest request, HttpServletResponse response)
            throws ServletException, IOException {
        processRequest(request, response);
    }
```

```
    /**
     * Returns a short description of the servlet.
     *
     * @return a String containing servlet description
     */
    @Override
    public String getServletInfo() {
        return "Short description";
    }// </editor-fold>

}
```

可见 Servlet 中的代码量很少，而且只有流程控制的代码。

10.8 DAO 设计模式总结

完成了 MIS 的第 5 版，就看到了 DAO 设计模式的全貌。图 10.9 展示了采用 DAO 设计模式之后的包结构。

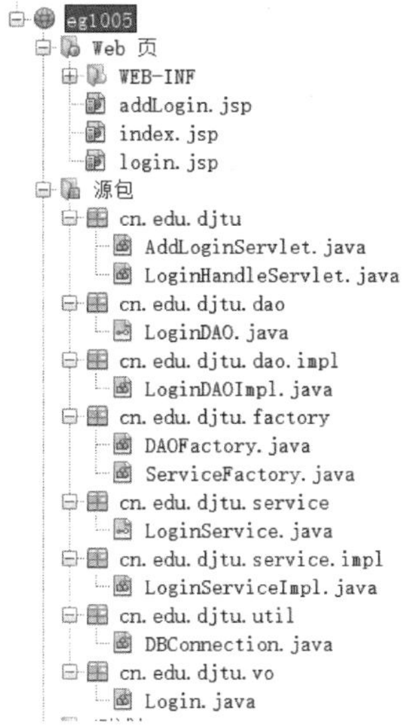

图 10.9 采用 DAO 设计模式后的包结构

对于简单的业务而言，DAO 设计模式的实现有如下 5 部分。

(1) 数据库连接类：连接数据库。

(2) POJO 实体类：封装数据。

(3) DAO 接口：定义对数据库中表的原子操作(增、删、改、查)。

(4) DAO 实现类：实现对数据库中表的原子操作(增、删、改、查)。

(5) DAO 工厂类：生产各种 DAO 实例。

此时，系统的层次结构如图 10.10 所示。

图 10.10　系统的层次结构图

10.9　本章回顾

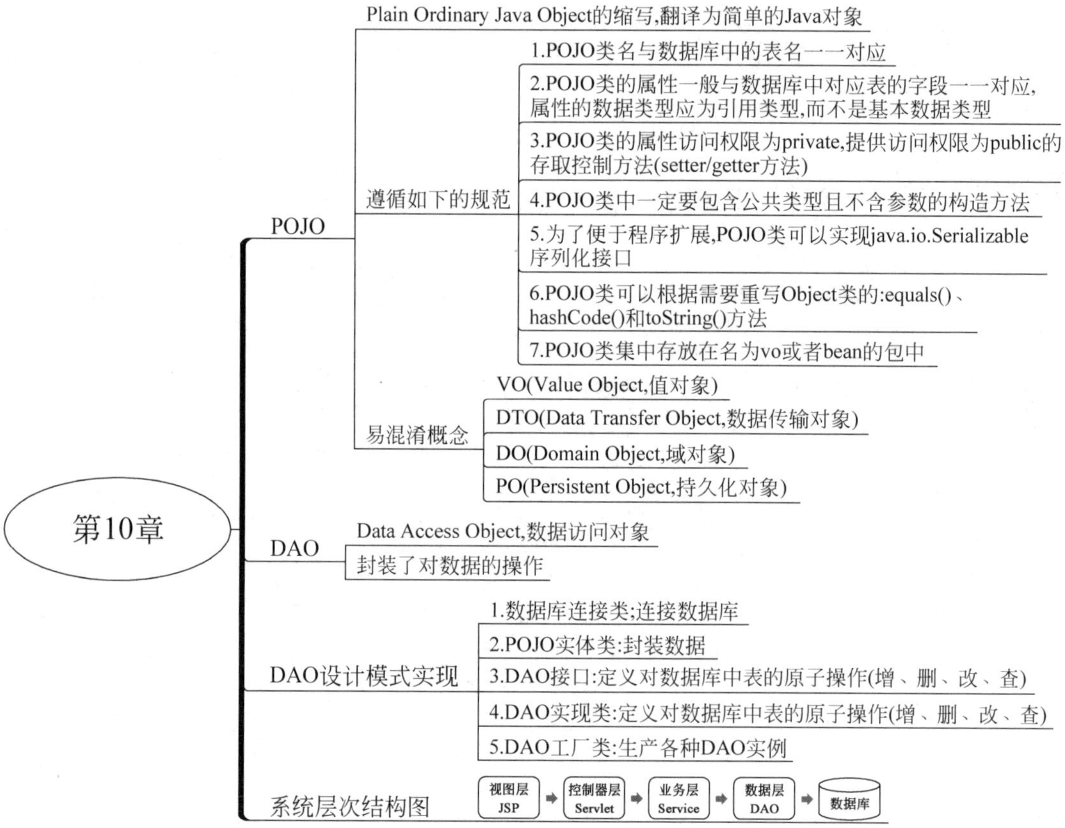

- 第10章
  - POJO
    - Plain Ordinary Java Object的缩写,翻译为简单的Java对象
    - 遵循如下的规范
      - 1.POJO类名与数据库中的表名一一对应
      - 2.POJO类的属性一般与数据库中对应表的字段一一对应,属性的数据类型应为引用类型,而不是基本数据类型
      - 3.POJO类的属性访问权限为private,提供访问权限为public的存取控制方法(setter/getter方法)
      - 4.POJO类中一定要包含公共类型且不含参数的构造方法
      - 5.为了便于程序扩展,POJO类可以实现java.io.Serializable序列化接口
      - 6.POJO类可以根据需要重写Object类的:equals()、hashCode()和toString()方法
      - 7.POJO类集中存放在名为vo或者bean的包中
    - 易混淆概念
      - VO(Value Object,值对象)
      - DTO(Data Transfer Object,数据传输对象)
      - DO(Domain Object,域对象)
      - PO(Persistent Object,持久化对象)
  - DAO
    - Data Access Object,数据访问对象
    - 封装了对数据的操作
  - DAO设计模式实现
    - 1.数据库连接类;连接数据库
    - 2.POJO实体类:封装数据
    - 3.DAO接口:定义对数据库中表的原子操作(增、删、改、查)
    - 4.DAO实现类:定义对数据库中表的原子操作(增、删、改、查)
    - 5.DAO工厂类:生产各种DAO实例
  - 系统层次结构图
    - 视图层 JSP → 控制器层 Servlet → 业务层 Service → 数据层 DAO → 数据库

图 10.11　第 10 章内容结构图

10.10　课后习题

1. 什么是 VO? Java EE 中一般用什么组件充当 VO。

2. 什么是 DAO? DAO 通常如何定义? 试举例说明。

3. 什么是 MVC? 在 Java EE 中,MVC 中的三部分通常可以由哪些组件充当?

4. 请在第 5 版实现的基础上,为项目添加修改密码的功能。修改密码操作的活动图如图 10.12 所示。

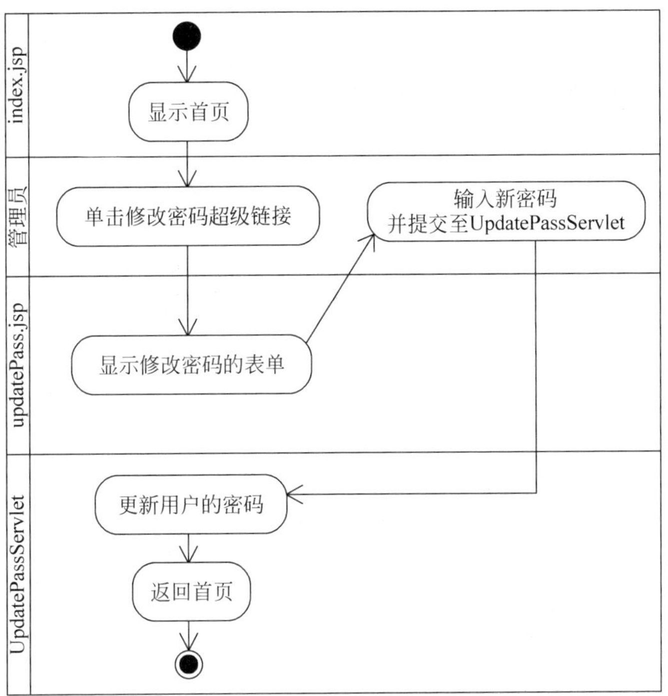

图 10.12 修改密码的活动图

第11章 客户信息管理系统(维护折扣码信息)

学习目标：

通过本章的学习，你应该：

- 熟练使用DAO设计模式完成项目

在第5章的学习中，曾经实现了折扣码信息维护的功能，在本章中将采用第10章掌握的技巧，实现新增折扣码、查看折扣码信息、更新折扣率、删除折扣码信息的功能。通过这个练习，进一步体会DAO设计模式。

11.1 系统用例图

在第10章的学习中，已经完成了登录和添加用户两个用例。在本章中，将继续完成新增折扣码、查看折扣码信息、更新折扣率、删除折扣码信息4个用例。系统用例图如图11.1所示。管理员登录系统后，在首页可以选择添加用户、新增折扣码、查看折扣码信息的操作。在查看折扣码信息的结果页面，可以完成更新折扣率和删除折扣码信息的操作。

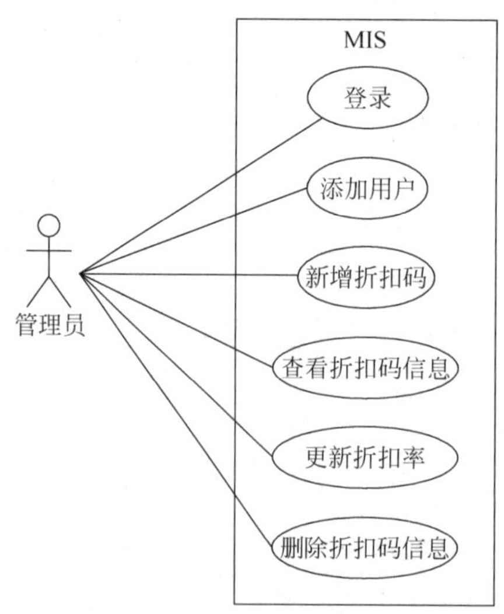

图11.1　MIS用例图

11.2 "新增折扣码"活动图

登录后管理员在首页单击"新增折扣码"的超级链接。显示新增折扣码的表单。管理员输入折扣码与折扣率信息并提交表单。如果该折扣码不存在,则新增折扣码成功并显示全部折扣码信息。否则提示该折扣码已存在,并显示新增折扣码表单。新增折扣码的活动图如图 11.2 所示。

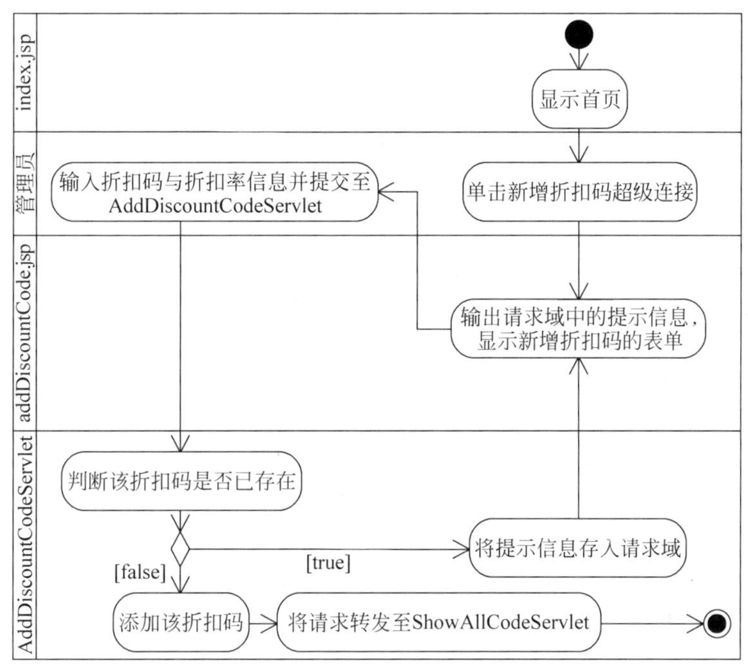

图 11.2 "新增折扣码"活动图

11.3 "查看全部折扣码信息"活动图

登录后管理员在首页单击"查看全部折扣码信息"的超级链接。如果折扣码信息存在,则显示全部折扣码信息,否则提示"暂无折扣码信息"。查看全部折扣码信息的活动图如图 11.3 所示。

11.4 "更新折扣率/删除折扣码信息"活动图

管理员在显示全部折扣码信息的页面可以选择"更新折扣率"或者"删除折扣码信息"的操作。

如果选择了"更新折扣率"的操作,则输入新的折扣率,单击"更新"按钮,更新成功后返回"显示折扣码信息"的页面,更新失败的话也返回"显示折扣码信息"的页面,并在页面提示

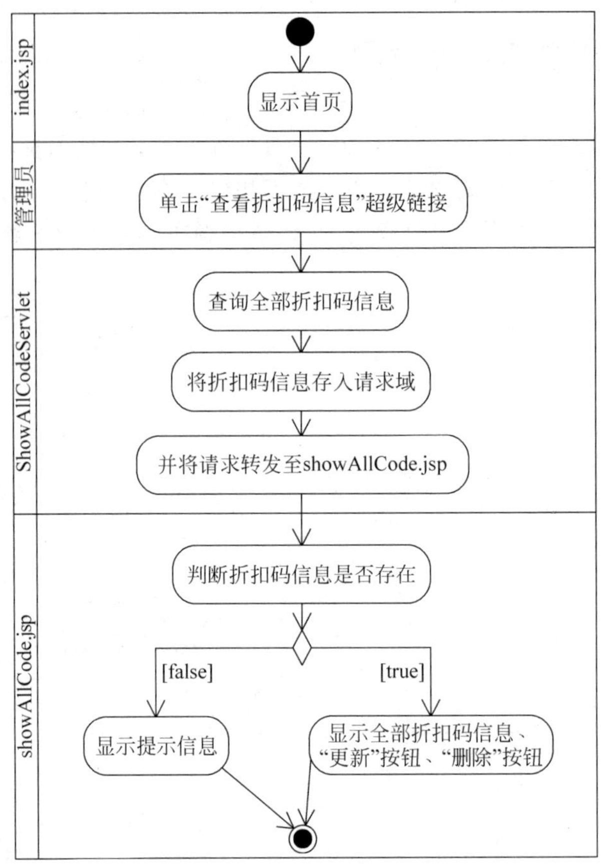

图 11.3 查询全部折扣码信息

"更新失败"。

如果选择了"删除折扣码"的操作，则单击"删除"按钮，删除成功后返回"显示折扣码信息"的页面，删除失败的话也返回"显示折扣码信息"的页面，并在页面提示"删除失败"。

"更新折扣率/删除折扣码信息"活动图如图 11.4 所示。

11.5 创建工程并编写 POJO 类代码、部分视图层代码

打开 NetBeans IDE，右击项目 eg1005，选择"复制"，在弹出的窗口中将项目名称设置为 eg1101，具体过程略。

工程名：eg1101。

依次创建如图 11.5 所示的文件，并将文件放置于正确的包内。(具体过程可参照第 10 章练习)

编写 DiscountCode.java 代码。DiscountCode 类与 sample 数据库中的 discount_code 表相对应。代码如下。

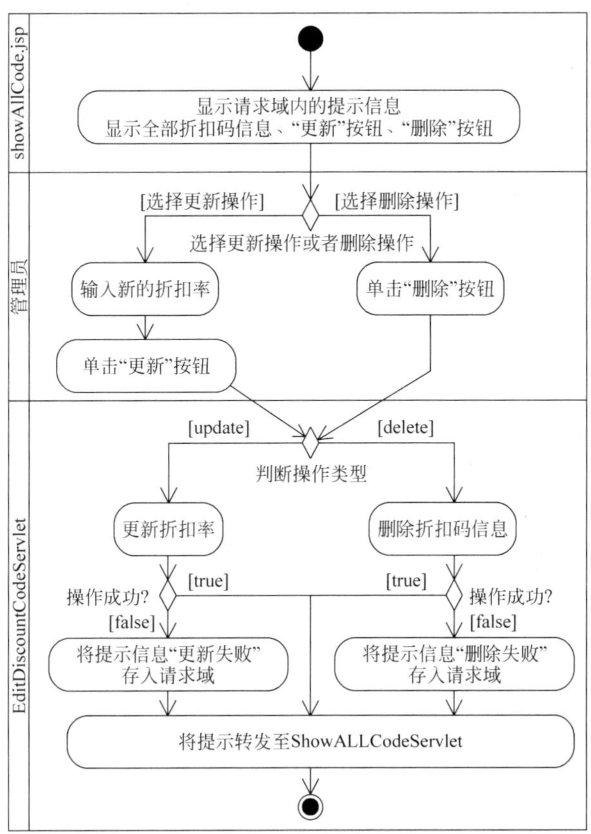

图 11.4　"更新折扣率/删除折扣码信息"活动图

图 11.5　eg1101 工程的文件结构

DiscountCode. java

```
package cn.edu.djtu.vo;

public class DiscountCode {
    private String discountCode;
    private Double rate;

    /**
     * @return the discountCode
     */
    public String getDiscountCode() {
        return discountCode;
    }

    /**
     * @param discountCode the discountCode to set
     */
    public void setDiscountCode(String discountCode) {
        this.discountCode = discountCode;
    }

    /**
     * @return the rate
     */
    public Double getRate() {
        return rate;
    }

    /**
     * @param rate the rate to set
     */
    public void setRate(Double rate) {
        this.rate = rate;
    }
}
```

编写首页 index. jsp 代码如下(为了代码简洁,此处未作样式控制):

index. jsp

```
<%@page contentType = "text/html" pageEncoding = "UTF-8" %>
<!DOCTYPE html>
<html>
    <head>
        <meta http-equiv = "Content-Type" content = "text/html; charset = UTF-8">
        <title>首页</title>
    </head>
    <body>
```

```
        <h1>Hello ${username}!</h1>
        <a href="addLogin.jsp">添加用户</a>||
        <a href="ShowAllCodeServlet">查看折扣码信息</a>||
        <a href="addDiscountCode.jsp">新增折扣码信息</a>
    </body>
</html>
```

编写 addDiscountCode.jsp 代码如下：

addDiscountCode.jsp

```
<%@page contentType="text/html" pageEncoding="UTF-8"%>
<!DOCTYPE html>
<html>
    <head>
        <meta http-equiv="Content-Type" content="text/html; charset=UTF-8">
        <title>新增折扣码</title>
    </head>
    <body>
        ${error}
        <form action="AddDiscountCodeServlet" method="post">
            <table align = "center">
                <tr>
                    <td>discount_code</td>
                    <td><input type="text" name="discount_code" /></td>
                </tr>
                <tr>
                    <td>rate</td><td><input type="text" name="rate" /></td>
                </tr>
                <tr>
                    <td></td><td><input type="submit" value="新增折扣码"/></td>
                </tr>
            </table>
        </form>
    </body>
</html>
```

11.6 编写 DAO 层代码

编写 DiscountCodeDAO.java 文件代码如下：

DiscountCodeDAO.java

```
package cn.edu.djtu.dao;

import cn.edu.djtu.vo.DiscountCode;
```

```
import java.util.List;

public interface DiscountCodeDAO {
    public abstract boolean save(DiscountCode discountCode) throws Exception;
    public abstract boolean delete(DiscountCode discountCode) throws Exception;
    public abstract boolean update(DiscountCode discountCode) throws Exception;
    public abstract DiscountCode findByCode(String code) throws Exception;
    public abstract List<DiscountCode> findAllCode() throws Exception;
}
```

编写 DiscountCodeDAOImpl.java 文件代码如下：

DiscountCodeDAOImpl.java

```
package cn.edu.djtu.dao.impl;

import cn.edu.djtu.dao.DiscountCodeDAO;
import cn.edu.djtu.vo.DiscountCode;
import java.sql.Connection;
import java.sql.PreparedStatement;
import java.sql.ResultSet;
import java.util.ArrayList;
import java.util.List;

public class DiscountCodeDAOImpl implements DiscountCodeDAO {

    private Connection conn = null;

    public DiscountCodeDAOImpl(Connection conn) {
        this.conn = conn;
    }

    @Override
    public boolean save(DiscountCode discountCode) throws Exception {
        boolean b = false;
        PreparedStatement prst;
        String sql = "INSERT INTO discount_code VALUES(?,?)";
        prst = conn.prepareStatement(sql);
        prst.setString(1, discountCode.getDiscountCode());
        prst.setDouble(2, discountCode.getRate());
        if (prst.executeUpdate() > 0) {
            b = true;
        }
        prst.close();
        return b;
    }

    @Override
    public boolean delete(DiscountCode discountCode) throws Exception {
```

```
        boolean b = false;
        PreparedStatement prst;
        String sql = "DELETE FROM discount_code WHERE discount_code = ?";
        prst = conn.prepareStatement(sql);
        prst.setString(1, discountCode.getDiscountCode());
        if (prst.executeUpdate() > 0) {
            b = true;
        }
        prst.close();
        return b;
    }

    @Override
    public boolean update(DiscountCode discountCode) throws Exception {
        boolean b = false;
        PreparedStatement prst;
        String sql = "update discount_code SET rate = ? WHERE discount_code = ?";
        prst = conn.prepareStatement(sql);
        prst.setDouble(1, discountCode.getRate());
        prst.setString(2, discountCode.getDiscountCode());
        if (prst.executeUpdate() > 0) {
            b = true;
        }
        prst.close();
        return b;
    }

    @Override
    public DiscountCode findByCode(String code) throws Exception {
        DiscountCode discountCode = null;
        PreparedStatement prst;
        ResultSet rs;
        String sql = "SELECT discount_code, rate FROM discount_code WHERE discount_code = ?";
        prst = conn.prepareStatement(sql);
        prst.setString(1, code);
        rs = prst.executeQuery();
        if (rs.next()) {
            DiscountCode dc = new DiscountCode();
            dc.setDiscountCode(rs.getString(1));
            dc.setRate(rs.getDouble(2));
        }
        rs.close();
        prst.close();
        return discountCode;
    }

    @Override
    public List<DiscountCode> findAllCode() throws Exception {
        List<DiscountCode> dcs = new ArrayList<>();
```

```
        PreparedStatement prst;
        ResultSet rs;
        String sql = "SELECT discount_code, rate FROM discount_code";
        prst = conn.prepareStatement(sql);
        rs = prst.executeQuery();
        while (rs.next()) {
            DiscountCode dc = new DiscountCode();
            dc.setDiscountCode(rs.getString(1));
            dc.setRate(rs.getDouble(2));
            dcs.add(dc);
        }
        rs.close();
        prst.close();
        return dcs;
    }
}
```

修改 DAOFactory.java 代码如下：

DAOFactory.java

```
public class DAOFactory {

  public static LoginDAOImpl getLoginDAOImpl(Connection conn) {
      return new LoginDAOImpl(conn);
  }

  public static DiscountCodeDAOImpl getDiscountCodeDAOImpl(Connection conn) {
      return new DiscountCodeDAOImpl(conn);
  }
}
```

11.7 编写 Service 层代码

编写 DiscountCodeService.java 文件代码如下：

DiscountCodeService.java

```
package cn.edu.djtu.service;

import cn.edu.djtu.vo.DiscountCode;
import java.util.List;

public interface DiscountCodeService {
    public abstract String addDiscountCode(DiscountCode dc);
    public abstract String removeDiscountCode(DiscountCode dc);
```

```
    public abstract String updateDiscountCode(DiscountCode dc);
    public abstract List<DiscountCode> showDiscountCode();
}
```

编写 DiscountCodeServiceImpl.java 文件代码如下：

DiscountCodeServiceImpl.java

```
package cn.edu.djtu.service.impl;

import cn.edu.djtu.dao.DiscountCodeDAO;
import cn.edu.djtu.factory.DAOFactory;
import cn.edu.djtu.service.DiscountCodeService;
import cn.edu.djtu.util.DBConnection;
import cn.edu.djtu.vo.DiscountCode;
import java.sql.Connection;
import java.util.List;
import java.util.logging.Level;
import java.util.logging.Logger;

public class DiscountCodeServiceImpl implements DiscountCodeService {

    private final DBConnection dbc = new DBConnection();

    @Override
    public String addDiscountCode(DiscountCode dc) {
        String string = "error";
        Connection conn = dbc.getConnection();
        DiscountCodeDAO dcdao = DAOFactory.getDiscountCodeDAOImpl(conn);
        try {
            if (dcdao.findByCode(dc.getDiscountCode()) == null && dcdao.save(dc)) {
                string = "success";
            }
        } catch (Exception ex) {
            Logger.getLogger(DiscountCodeServiceImpl.class.getName()).log(Level.SEVERE,
null, ex);
        } finally {
            dbc.close();
        }
        return string;
    }

    @Override
    public String removeDiscountCode(DiscountCode dc) {
        String string = "error";
        Connection conn = dbc.getConnection();
        DiscountCodeDAO dcdao = DAOFactory.getDiscountCodeDAOImpl(conn);
        try {
            if (dcdao.delete(dc)) {
```

```
                string = "success";
            }
        } catch (Exception ex) {
            Logger.getLogger(DiscountCodeServiceImpl.class.getName()).log(Level.SEVERE,
null, ex);
        } finally {
            dbc.close();
        }
        return string;
    }

    @Override
    public String updateDiscountCode(DiscountCode dc) {
        String string = "error";
        Connection conn = dbc.getConnection();
        DiscountCodeDAO dcdao = DAOFactory.getDiscountCodeDAOImpl(conn);
        try {
            if (dcdao.update(dc)) {
                string = "success";
            }
        } catch (Exception ex) {
            Logger.getLogger(DiscountCodeServiceImpl.class.getName()).log(Level.SEVERE,
null, ex);
        } finally {
            dbc.close();
        }
        return string;
    }

    @Override
    public List<DiscountCode> showDiscountCode() {
        List<DiscountCode> dcs = null;
        Connection conn = dbc.getConnection();
        DiscountCodeDAO dcdao = DAOFactory.getDiscountCodeDAOImpl(conn);
        try {
            dcs = dcdao.findAllCode();
        } catch (Exception ex) {
            Logger.getLogger(DiscountCodeServiceImpl.class.getName()).log(Level.SEVERE,
null, ex);
        } finally {
            dbc.close();
        }
        return dcs;
    }
}
```

修改 ServiceFactory.java 文件代码如下：

ServiceFactory.java

```
package cn.edu.djtu.factory;

import cn.edu.djtu.service.impl.DiscountCodeServiceImpl;
import cn.edu.djtu.service.impl.LoginServiceImpl;

public class ServiceFactory {

    public static LoginServiceImpl getLoginServiceImpl() {
        return new LoginServiceImpl();
    }

    public static DiscountCodeServiceImpl getDiscountCodeServiceImpl() {
        return new DiscountCodeServiceImpl();
    }
}
```

11.8 编写控制器层代码

编写 ShowAllCodeServlet.java 文件代码如下：

ShowAllCodeServlet.java

```
package cn.edu.djtu;

import cn.edu.djtu.factory.ServiceFactory;
import cn.edu.djtu.service.DiscountCodeService;
import java.io.IOException;
import javax.servlet.ServletException;
import javax.servlet.annotation.WebServlet;
import javax.servlet.http.HttpServlet;
import javax.servlet.http.HttpServletRequest;
import javax.servlet.http.HttpServletResponse;

@WebServlet(name = "ShowAllCodeServlet", urlPatterns = {"/ShowAllCodeServlet"})
public class ShowAllCodeServlet extends HttpServlet {

    /**
     * Processes requests for both HTTP <code>GET</code> and <code>POST</code>
     * methods.
     *
     * @param request servlet request
     * @param response servlet response
     * @throws ServletException if a servlet-specific error occurs
     * @throws IOException if an I/O error occurs
     */
    protected void processRequest(HttpServletRequest request, HttpServletResponse response)
            throws ServletException, IOException {
```

```
        DiscountCodeService dcs = ServiceFactory.getDiscountCodeServiceImpl();
        request.setAttribute("list", dcs.showDiscountCode());
        request.getRequestDispatcher("showAllCode.jsp").forward(request, response);
    }

    // <editor-fold defaultstate="collapsed" desc="HttpServlet methods. Click on the +
sign on the left to edit the code.">
    /**
     * Handles the HTTP <code>GET</code> method.
     *
     * @param request servlet request
     * @param response servlet response
     * @throws ServletException if a servlet-specific error occurs
     * @throws IOException if an I/O error occurs
     */
    @Override
    protected void doGet(HttpServletRequest request, HttpServletResponse response)
            throws ServletException, IOException {
        processRequest(request, response);
    }

    /**
     * Handles the HTTP <code>POST</code> method.
     *
     * @param request servlet request
     * @param response servlet response
     * @throws ServletException if a servlet-specific error occurs
     * @throws IOException if an I/O error occurs
     */
    @Override
    protected void doPost(HttpServletRequest request, HttpServletResponse response)
            throws ServletException, IOException {
        processRequest(request, response);
    }

    /**
     * Returns a short description of the servlet.
     *
     * @return a String containing servlet description
     */
    @Override
    public String getServletInfo() {
        return "Short description";
    }// </editor-fold>

}
```

编写 AddDiscountCodeServlet. java 文件代码如下：

AddDiscountCodeServlet. java

```
package cn.edu.djtu;

import cn.edu.djtu.factory.ServiceFactory;
import cn.edu.djtu.service.DiscountCodeService;
import cn.edu.djtu.vo.DiscountCode;
import java.io.IOException;
import javax.servlet.ServletException;
import javax.servlet.annotation.WebServlet;
import javax.servlet.http.HttpServlet;
import javax.servlet.http.HttpServletRequest;
import javax.servlet.http.HttpServletResponse;

@WebServlet(name = "AddDiscountCodeServlet", urlPatterns = {"/AddDiscountCodeServlet"})
public class AddDiscountCodeServlet extends HttpServlet {

    /**
     * Processes requests for both HTTP <code>GET</code> and <code>POST</code>
     * methods.
     *
     * @param request servlet request
     * @param response servlet response
     * @throws ServletException if a servlet-specific error occurs
     * @throws IOException if an I/O error occurs
     */
    protected void processRequest(HttpServletRequest request, HttpServletResponse response)
            throws ServletException, IOException {
        String discount_code = request.getParameter("discount_code");
        String rate = request.getParameter("rate");

        DiscountCode dc = new DiscountCode();
        dc.setDiscountCode(discount_code);
        dc.setRate(Double.parseDouble(rate));

        DiscountCodeService dcs = ServiceFactory.getDiscountCodeServiceImpl();
        if (dcs.addDiscountCode(dc).equals("error")) {
            request.setAttribute("error", "该折扣码已存在");
            request.getRequestDispatcher("addDiscountCode.jsp").forward(request,
response);
        } else {
            request.getRequestDispatcher("ShowAllCodeServlet").forward(request,
response);
        }

    }
```

```
    // <editor-fold defaultstate="collapsed" desc="HttpServlet methods. Click on the +
sign on the left to edit the code.">
    /**
     * Handles the HTTP <code>GET</code> method.
     *
     * @param request servlet request
     * @param response servlet response
     * @throws ServletException if a servlet-specific error occurs
     * @throws IOException if an I/O error occurs
     */
    @Override
    protected void doGet(HttpServletRequest request, HttpServletResponse response)
            throws ServletException, IOException {
        processRequest(request, response);
    }

    /**
     * Handles the HTTP <code>POST</code> method.
     *
     * @param request servlet request
     * @param response servlet response
     * @throws ServletException if a servlet-specific error occurs
     * @throws IOException if an I/O error occurs
     */
    @Override
    protected void doPost(HttpServletRequest request, HttpServletResponse response)
            throws ServletException, IOException {
        processRequest(request, response);
    }

    /**
     * Returns a short description of the servlet.
     *
     * @return a String containing servlet description
     */
    @Override
    public String getServletInfo() {
        return "Short description";
    }// </editor-fold>

}
```

编写 EditDiscountCodeServlet.java 文件代码如下：

EditDiscountCodeServlet.java

```
package cn.edu.djtu;

import cn.edu.djtu.factory.ServiceFactory;
```

```
import cn.edu.djtu.service.DiscountCodeService;
import cn.edu.djtu.vo.DiscountCode;
import java.io.IOException;
import javax.servlet.ServletException;
import javax.servlet.annotation.WebServlet;
import javax.servlet.http.HttpServlet;
import javax.servlet.http.HttpServletRequest;
import javax.servlet.http.HttpServletResponse;

@WebServlet(name = "EditDiscountCodeServlet", urlPatterns = {"/EditDiscountCodeServlet"})
public class EditDiscountCodeServlet extends HttpServlet {

    /**
     * Processes requests for both HTTP <code>GET</code> and <code>POST</code>
     * methods.
     *
     * @param request servlet request
     * @param response servlet response
     * @throws ServletException if a servlet-specific error occurs
     * @throws IOException if an I/O error occurs
     */
    protected void processRequest(HttpServletRequest request, HttpServletResponse response)
            throws ServletException, IOException {
        String discount_code = request.getParameter("discount_code");
        String rate = request.getParameter("rate");
        String submit = request.getParameter("submit");

        DiscountCode dc = new DiscountCode();
        dc.setDiscountCode(discount_code);
        dc.setRate(Double.parseDouble(rate));

        DiscountCodeService dcs = ServiceFactory.getDiscountCodeServiceImpl();
        if (submit.equals("update")) {
            if (dcs.updateDiscountCode(dc).equals("error")) {
                request.setAttribute("error", "更新失败");
            }
        } else {
            if (dcs.removeDiscountCode(dc).equals("error")) {
                request.setAttribute("error", "删除失败");
            }
        }
        request.getRequestDispatcher("ShowAllCodeServlet").forward(request, response);
    }

    // <editor-fold defaultstate="collapsed" desc="HttpServlet methods. Click on the + sign on the left to edit the code.">
    /**
     * Handles the HTTP <code>GET</code> method.
```

```
     *
     * @param request servlet request
     * @param response servlet response
     * @throws ServletException if a servlet - specific error occurs
     * @throws IOException if an I/O error occurs
     */
    @Override
    protected void doGet(HttpServletRequest request, HttpServletResponse response)
            throws ServletException, IOException {
        processRequest(request, response);
    }

    /**
     * Handles the HTTP <code>POST</code> method.
     *
     * @param request servlet request
     * @param response servlet response
     * @throws ServletException if a servlet - specific error occurs
     * @throws IOException if an I/O error occurs
     */
    @Override
    protected void doPost(HttpServletRequest request, HttpServletResponse response)
            throws ServletException, IOException {
        processRequest(request, response);
    }

    /**
     * Returns a short description of the servlet.
     *
     * @return a String containing servlet description
     */
    @Override
    public String getServletInfo() {
        return "Short description";
    }// </editor - fold>

}
```

11.9 编写其他视图层代码

编写首页 showAllcode.jsp 代码如下(为了代码简洁,此处未做样式控制):

showAllcode.jsp

```
<%@taglib prefix = "c" uri = "http://java.sun.com/jsp/jstl/core" %>
<%@page contentType = "text/html" pageEncoding = "UTF - 8" %>
<!DOCTYPE html>
```

```
<html>
  <head>
    <meta http-equiv="Content-Type" content="text/html; charset=UTF-8">
    <title>JSP Page</title>
  </head>
  <body>
     ${error}
    <c:choose>
      <c:when test="${empty list}">
        暂无折扣码信息<a href="addDiscountCode.jsp">新增折扣码</a>
      </c:when>
      <c:otherwise>
        <table border = "1" align = "center">
            <caption>折扣码信息</caption>
            <tr>
                <th>discount_code</th>
                <th>rate</th>
                <th>更新折扣率</th>
                <th>删除折扣码</th>
            </tr>
        <c:forEach var="code" items="${list}">
            <form action="EditDiscountCodeServlet" method="post">
            <tr>
                <td>
                    ${code.discountCode}
        <input type="hidden" name="discount_code" value="${code.discountCode}"/>
                </td>
                <td>
                    <input type="text" name="rate" value="${code.rate}"/>
                </td>
                <td>
                    <input type="submit" name="submit" value="update"/>
                </td>
                <td>
                    <input type="submit" name="submit" value="delete"/>
                </td>
            </tr>
            </form>
          </c:forEach>
          </table>
      </c:otherwise>
    </c:choose>
    </body>
</html>
```

11.10 系统目前存在的问题

首先，本章的案例侧重于学习 DAO 设计模式，偏重于服务器端编程，对于必要的前端校验均未进行。例如：在新增折扣码时，如果折扣码和折扣率未录入就提交的话，会产生错误码为 500 的错误，如图 11.6 所示。这部分校验一般由前端工程师使用 JS(Java Script)来解决。

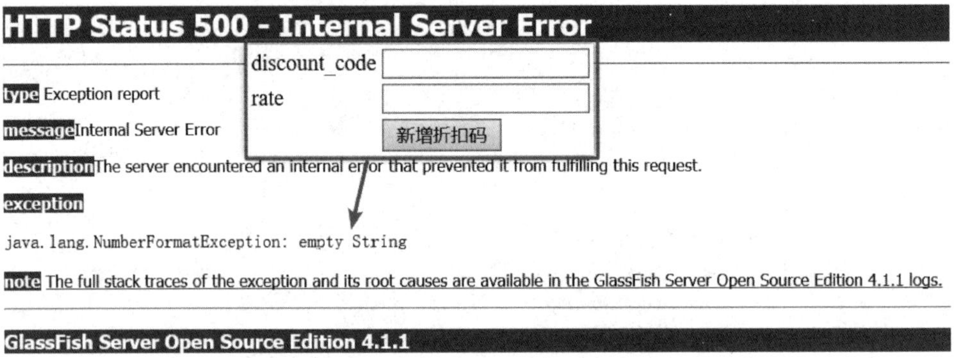

HTTP Status 500 - Internal Server Error

discount_code

rate

新增折扣码

type Exception report

message Internal Server Error

description The server encountered an internal error that prevented it from fulfilling this request.

exception

java.lang.NumberFormatException: empty String

note The full stack traces of the exception and its root causes are available in the GlassFish Server Open Source Edition 4.1.1 logs.

GlassFish Server Open Source Edition 4.1.1

图 11.6 输入未经校验的数据后抛出的异常

其次，为了代码尽可能简洁，便于介绍和学习，没有做页面美化的工作。这部分工作一般由前端工程师使用 CSS(Cascading Style Sheets，层叠样式表)来解决。

第12章 数据库访问技术补足

学习目标：

通过本章的学习，你应该：

- 熟练掌握读取属性文件的方法
- 熟练掌握通过数据库连接池访问数据库的方法

在第 11 章的案例中，编写了数据库连接类 DBConnection，将获得数据库连接对象和关闭数据库连接对象这两个操作集中封装在一起。但是观察如下代码可知，当数据库参数需要改变时，依然需要修改 DBConnection 类的源代码。

```
public class DBConnection {
    private static final String URL = "jdbc:derby://localhost:1527/sample";
    private static final String DRIVER = "org.apache.derby.jdbc.ClientDriver";
    private static final String USER = "app";
    private static final String PASSWORD = "app";
    //其他代码略
}
```

那么是否能做到即使参数改变，也不需要更改 DBConnection 类的源代码呢？本章将学习两种方法解决这个问题。

(1) 将参数写入属性文件，在程序中读取参数。

(2) 使用数据库连接池。

12.1 读取属性文件中的数据库配置信息

在 Web 应用程序中可以创建属性文件，属性文件很像文本文件，可以使用文本编辑器打开，方便地编辑。即使改变了属性文件，也不需要重新编译 DBConnection. java 文件，只要重启服务器，在程序中重新读取就可以了。如果将数据库访问的配置信息写在属性文件里，那么一旦需要做数据库迁移之类的操作时，只要用文本编辑器修改属性文件就可以了。

属性文件的后缀是. properties，例如：可以命名（写有数据库配置信息属性文件）为 database. properties。读取属性文件有很多种方式，本章主要应用 Properties 类。总的编程思路是这样的：首先在 DBConnection. java 文件所在目录新建一个属性文件，里面存有数据库访问的参数信息。然后改写之前的 DBConnection 类，读取参数信息。

打开 NetBeans IDE,右击项目 eg1101,选择“复制”命令,在弹出的窗口中将项目名称设置为 eg1201,具体过程略。

工程名：eg1201。

右击 cn.edu.djtu.util 包,选择“新建”→“其他”命令,如图 12.1 所示。

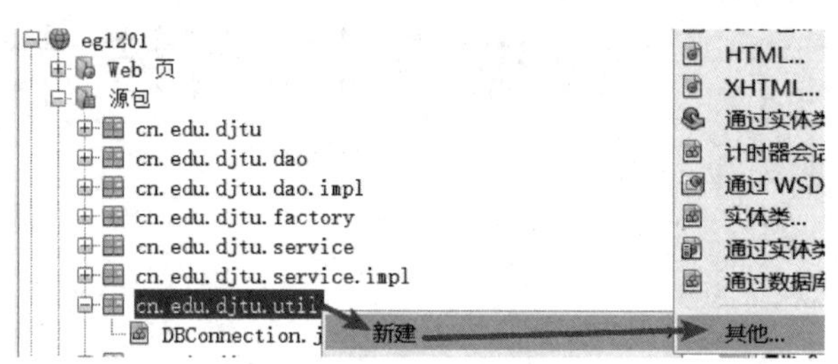

图 12.1　新建属性文件步骤 1

在弹出的窗口中选择“类别”中的“其他”,“文件类型”选择“属性文件”,单击“下一步”按钮,如图 12.2 所示。

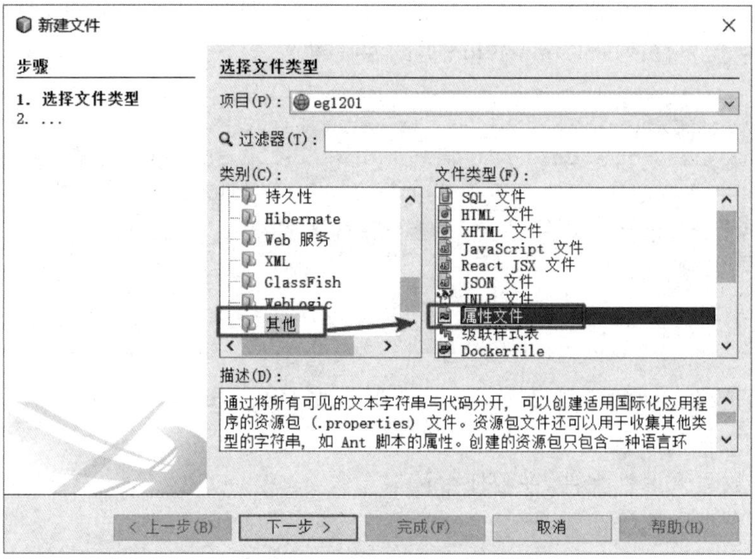

图 12.2　新建属性文件步骤 2

在弹出的窗体中输入文件名为 database,单击“完成”按钮,至此属性文件创建完毕,如图 12.3 所示。

编写 database.properties 代码如下,信息以 key-value 对应的形式存放。

database.properties

```
driver = org.apache.derby.jdbc.ClientDriver
url = jdbc:derby://localhost:1527/sample
user = app
password = app
```

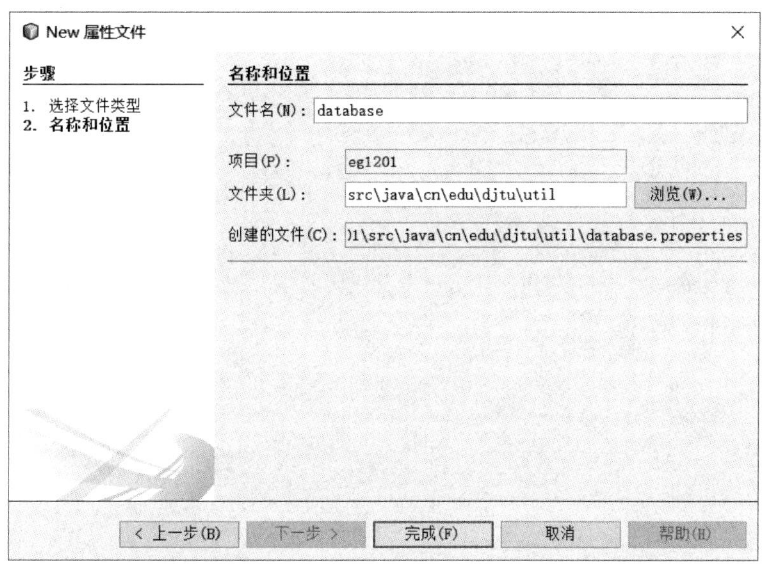

图 12.3　新建属性文件步骤 3

改写 DBConnection.java 代码如下,不同的地方见波浪线部分(其中一个变化是去掉了参数变量前面的 final,因为参数值需要在程序中读入)。

DBConnection.java

```
package cn.edu.djtu.util;

import java.io.IOException;
import java.io.InputStream;
import java.sql.Connection;
import java.sql.DriverManager;
import java.sql.SQLException;
import java.util.Properties;
import java.util.logging.Level;
import java.util.logging.Logger;

public class DBConnection {
    private static String URL = "";
    private static String DRIVER = "";
    private static String USER = "";
    private static String PASSWORD = "";
    private Connection conn = null;

    static {
    try (InputStream is = DBConnection.class.getResourceAsStream("database.properties");){
            Properties properties = new Properties();
            properties.load(is);
            URL = properties.getProperty("url");
```

```
                DRIVER = properties.getProperty("driver");
                USER = properties.getProperty("user");
                PASSWORD = properties.getProperty("password");
            } catch (IOException ex) {
                Logger.getLogger(DBConnection.class.getName()).log(Level.SEVERE, null, ex);
            }
        }

        public DBConnection() {
            try {
                Class.forName(DRIVER);
                conn = DriverManager.getConnection(URL, USER, PASSWORD);
            } catch (ClassNotFoundException | SQLException ex) {
    Logger.getLogger(DBConnection.class.getName()).log(Level.SEVERE, null, ex);
            }
        }
        public Connection getConnection(){
            return conn;
        }
        public void close(){
            try {
                conn.close();
            } catch (SQLException ex) {
                Logger.getLogger(DBConnection.class.getName()).log(Level.SEVERE, null, ex);
            }
        }
    }
```

12.2 采用数据库连接池访问数据库

使用JDBC访问数据库的过程中，无论是读数据库，还是写数据库，都要涉及建立连接和关闭连接，这样的操作都会消耗一定的资源。在访问量不大的情况下，这样的技术方案没有什么问题，但是如果访问量比较大时，那么频繁的创建对象、销毁对象，将会导致系统性能急剧下降。而连接池可以在一定程度上解决这一问题。

连接池的基本思想是在系统初始化的时候，将数据库连接作为对象存储在内存中，当用户需要访问数据库时，不是建立一个新的连接，而是从连接池中取出一个已建立的空闲连接对象。使用完毕后，用户也不是将连接关闭，而是将连接放回连接池中，以供下一个请求访问使用。连接的建立和断开都由连接池自身来管理。同时，还可以通过设置连接池的参数来控制连接池中的初始连接数和连接的上下限数以及每个连接的最大使用次数、最大空闲时间等。也可以通过其自身的管理机制来监视数据库连接的数量和使用情况等。

选择“服务”选项卡的“服务器”选项，可以看到GlassFish Server已经在资源里JDBC菜单下配置好了一个针对Sample数据库的数据库连接池SamplePool，JDBC资源名为jdbc/sample，可以即刻使用，如图12.4所示。

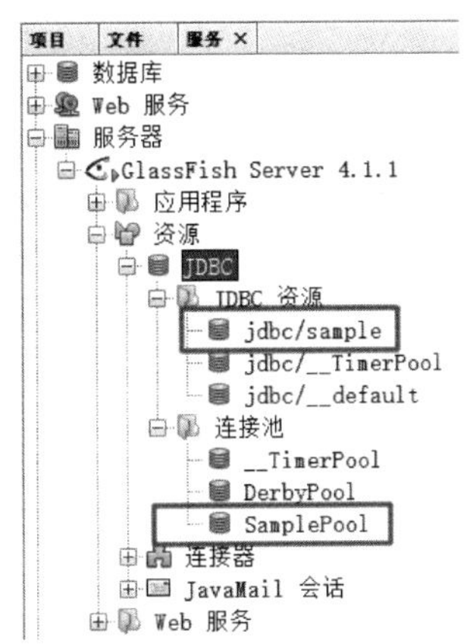

图 12.4 针对 Sample 数据库的数据库连接池

打开 NetBeans IDE,右击项目 eg1101,选择"复制"命令,在弹出的窗口中将项目名称设置为 eg1202,具体过程略。

工程名: eg1202。

改写 DBConnection.java 代码如下所示,变化的地方如波浪线所示。

DBConnection.java

```
package cn.edu.djtu.util;

import java.sql.Connection;
import java.sql.SQLException;
import java.util.logging.Level;
import java.util.logging.Logger;
import javax.naming.InitialContext;
import javax.naming.NamingException;
import javax.sql.DataSource;

public class DBConnection {
    private Connection conn = null;
    public DBConnection() {
        try {
            InitialContext initialContext = new InitialContext();
            DataSource ds = (DataSource) initialContext.lookup("jdbc/sample");
            conn = ds.getConnection();
        } catch (NamingException | SQLException ex) {
    Logger.getLogger(DBConnection.class.getName()).log(Level.SEVERE, null, ex);
        }
}

    public Connection getConnection() {
        return conn;
    }
```

```
    public void close() {
        try {
            conn.close();
        } catch (SQLException ex) {
            Logger.getLogger(DBConnection.class.getName()).log(Level.SEVERE, null, ex);
        }
    }
}
```

项目其他代码无须改变即可完成原来的功能，由此也可见分层的程序架构，当某一层改变时，只要保持接口不变，那么层与层之间互不影响。

在本次练习时，采用了现成的数据库连接池。如果要自己配置数据库连接池，可以参照相关开发环境的使用手册或帮助文档。

12.3 访问其他数据库

通过本章的讲解，大家对数据的访问，从认识上应该又加深了一层，现在来探讨一下如何访问其他数据库。

访问任何一个数据库，需要知道的信息都差不多，主要是 4 个参数。例如，在逻辑结构不变的情况下，如果想要把数据库从 JavaDB 迁移到 MySQL，那么可能访问 MySQL 数据库的属性文件也许是这样的：

database. properties

```
driver = com.mysql.jdbc.Driver
url = jdbc:mysql://localhost:3306/sample?useUnicode = true&charset = uf-8
user = root
password = 111111
```

可见只要把 driver、url、username、password 设置好就可以了。访问 Oracle 或者 SQL Server 也是类似的。

12.4 课后习题

1. 完成本章的样例，比较各样例之间的细微差异。
2. 尝试用属性文件方式访问自己熟悉的数据库。
3. 尝试用数据库连接池方式访问自己熟悉的数据。

参考文献

[1] Marty Hall, Larry Brown. Servlet 与 JSP 核心编程(第二版)[M]. 赵学良,译. 北京:清华大学出版社,2004.

[2] 孙鑫. Java Web 开发详解[M]. 北京:电子工业出版社,2006.

[3] 巴萨姆,等. HeadFirstServlets&JSP(中文版)[M]. 苏钰函,林剑,译. 北京:中国电力出版社,2006.

[4] 郝玉龙,姜桦. Java EE 编程技术[M]. 北京:清华大学出版社,北京交通大学出版社,2008.

[5] 缪勇,等. JSP 网络开发逐步深入[M]. 北京:清华大学出版社,2010.

[6] 林培光,耿长欣,张燕. Java EE 简明教程[M]. 北京:清华大学出版社,2012.

[7] 王晓华,李丹程,徐洪智. Java EE 架构与程序设计[M]. 北京:电子工业出版社,2011.

[8] 俞东进,任祖杰. Java EE Web 应用开发[M]. 北京:电子工业出版社,2012.

[9] 覃华,韦兆文,陈琴. JSP 2.0 大学教程[M]. 北京:机械工业出版社,2008.

[10] 郭真,王国辉. JSP 程序设计教程[M]. 北京:人民邮电出版社,2010.

电子资源

[1] 百度百科. Java EE [EB/OL]. (2018-07-04) [2018-12-07]. https://baike. baidu. com/item/Java%20EE.

[2] 百度百科. HTML[EB/OL]. (2018-11-20)[2018-12-07]. https://baike. baidu. com/item/HTML.

[3] w3school 在线教程. HTML/CSS 教程[EB/OL]. (2018-12-07)[2018-12-07]. http://www. w3school. com. cn/h. asp.

[4] 百度百科. Servlet[EB/OL]. (2018-06-26)[2018-12-07]. https://baike. baidu. com/item/servlet.

[5] w3school 在线教程. MIME 参考手册[EB/OL]. (2018-12-07)[2018-12-07]. http://www. w3school. com. cn/media/media_mimeref. asp.

[6] 百度百科. Cookie [EB/OL]. (2018-09-05) [2018-12-07]. https://baike. baidu. com/item/cookie/1119.

[7] 百度百科. JDBC[EB/OL]. (2017-06-06)[2018-12-07] . https://baike. baidu. com/item/jdbc.

[8] 百度百科. JSP[EB/OL]. (2018-06-06)[2018-12-07]. https://baike. baidu. com/item/JSP/141543.

[9] 百度百科. 数据库连接池[EB/OL]. (2016-04-15)[2018-12-07] . https://baike. baidu. com/item/数据库连接池.

[10] 林信良. JSP Tutorial [EB/OL]. (2018-12-07)[2018-12-07]. http://www. tutorialspoint. com/jsp/index. htm.

图书资源支持

感谢您一直以来对清华版图书的支持和爱护。为了配合本书的使用，本书提供配套的资源，有需求的读者请扫描下方的“书圈”微信公众号二维码，在图书专区下载，也可以拨打电话或发送电子邮件咨询。

如果您在使用本书的过程中遇到了什么问题，或者有相关图书出版计划，也请您发邮件告诉我们，以便我们更好地为您服务。

我们的联系方式：

地　　址：北京市海淀区双清路学研大厦 A 座 701

邮　　编：100084

电　　话：010-62770175-4608

资源下载：http://www.tup.com.cn

客服邮箱：tupjsj@vip.163.com

QQ：2301891038（请写明您的单位和姓名）

资源下载、样书申请

书圈

扫一扫，获取最新目录

用微信扫一扫右边的二维码，即可关注清华大学出版社公众号“书圈”。